A Primer on Pseudorandom Generators

University
LECTURE
Series

Volume 55

A Primer on Pseudorandom Generators

Oded Goldreich

American Mathematical Society
Providence, Rhode Island

2010 *Mathematics Subject Classification.* Primary 68-01, 68-02, 68Q01, 68R01;
Secondary 68Q15, 68Q17, 68W20.

For additional information and updates on this book, visit
www.ams.org/bookpages/ulect-55

Library of Congress Cataloging-in-Publication Data

Goldreich, Oded.
A primer on pseudorandom generators / Oded Goldreich.
p. cm. — (University lecture series ; v. 55)
Includes bibliographical references and index.
ISBN 978-0-8218-5192-0 (alk. paper)
1. Computational complexity. 2. Random number generators. 3. Computer science–Mathematics. I. Title

QA267.7.G654 2010
004.01′51–dc22 2010018152

∞ The paper used in this book is acid-free and falls within the guidelines
established to ensure permanence and durability.
Visit the AMS home page at http://www.ams.org/

10 9 8 7 6 5 4 3 2 1 15 14 13 12 11 10

Contents

Preface

Indistinguishable things are identical.[1]

G.W. Leibniz (1646–1714)

This primer to the theory of pseudorandomness presents a fresh look at the *question of randomness*, which arises from a complexity theoretic approach to randomness. The crux of this (complexity theoretic) approach is the postulate that a distribution is random (or rather pseudorandom) if it cannot be distinguished from the uniform distribution by *any efficient procedure.* Thus, (pseudo)randomness is not an inherent property of an object, but is rather subjective to the observer.

At the extreme, this approach says that the question of whether the world is actually deterministic or allows for some free choice (which may be viewed as a source of randomness) is irrelevant. *What matters is how the world looks to us and to various computationally bounded devices.* That is, if some phenomenon looks random, then we may treat it as if it is random. Likewise, if we can generate sequences that cannot be distinguished from the uniform distribution by any efficient procedure, then we can use these sequences in any efficient randomized application instead of the ideal coin tosses that are postulated in the design of this application.

The pivot of the foregoing approach is the notion of *computational indistinguishability*, which refers to pairs of distributions that cannot be distinguished by efficient procedures. The most fundamental incarnation of this notion associates efficient procedures with polynomial-time algorithms, but other incarnations that restrict attention to different classes of distinguishing procedures also lead to important insights. Likewise, the *effective generation* of pseudorandom objects, which is of major concern, is actually a general paradigm with numerous useful incarnations (which differ in the computational complexity limitations imposed on the generation process).

Following the foregoing principles, we briefly outline some of the key elements of the theory of pseudorandomness. Indeed, the key concept is that of a pseudorandom generator, which is an efficient deterministic procedure that stretches short random seeds into longer pseudorandom sequences. Thus, a generic formulation of pseudorandom generators consists of specifying three fundamental aspects – the *stretch measure* of the generators; the class of distinguishers that the generators are

[1] This is Leibniz's *Principle of Identity of Indiscernibles.* Leibniz admits that counterexamples to this principle are conceivable but will not occur in real life because God is much too benevolent. We thus believe that he would have agreed to the theme of this text, which asserts that *indistinguishable things should be considered as if they were identical.*

supposed to fool (i.e., the algorithms with respect to which the *computational indistinguishability* requirement should hold); and the resources that the generators are allowed to use (i.e., their own *computational complexity*).

The archetypical case of pseudorandom generators refers to efficient generators that fool any feasible procedure; that is, the potential distinguisher is any probabilistic polynomial-time algorithm, which may be more complex than the generator itself (which, in turn, has time-complexity bounded by a fixed polynomial). These generators are called general-purpose, because their output can be safely used in any efficient application. Such (general-purpose) pseudorandom generators exist if and only if there exist functions (called one-way functions) that are easy to evaluate but hard to invert.

In contrast to such (general-purpose) pseudorandom generators, for the purpose of derandomization (i.e., converting randomized algorithms into corresponding deterministic ones), a relaxed definition of pseudorandom generators suffices. In particular, for such a purpose, one may use pseudorandom generators that are somewhat more complex than the potential distinguisher (which represents a randomized algorithm to be derandomized). Following this approach, adequate pseudorandom generators yield a full derandomization of probabilistic polynomial-time algorithms (e.g., $\mathcal{BPP} = \mathcal{P}$), and such generators can be constructed based on the assumption that some exponential-time solvable problems (i.e., problems in $\mathcal{E}$) have no sub-exponential size circuits.

Indeed, both the general-purpose pseudorandom generators and the aforementioned "derandomizers" demonstrate that randomness and computational difficulty are related. This trade-off is not surprising in light of the fact that the very definition of pseudorandomness refers to computational difficulty (i.e., the difficulty of distinguishing the pseudorandom distribution from a truly random one).

Finally, we mention that it is also beneficial to consider pseudorandom generators that fool space-bounded distinguishers and generators that exhibit some limited random behavior (e.g., outputting a pairwise independent or a small-bias sequence). Such (special-purpose) pseudorandom generators can be constructed without relying on any computational complexity assumptions, because the behavior of the corresponding (limited) distinguishers can be analyzed even at the current historical time. Nevertheless, such (special-purpose) pseudorandom generators offer numerous applications.

Note: The study of pseudorandom generators is part of complexity theory (cf. e.g., [24]), and some basic familiarity with complexity theory will be assumed in the current text. In fact, the current primer is an abbreviated (and somewhat revised) version of [24, Chap. 8]. Nevertheless, we believe that there are merits to providing a separate treatment of the theory of pseudorandomness, since this theory is of natural interest to various branches of mathematics and science. In particular, we hope to reach readers that may not have a general interest in complexity theory at large and/or do not wish to purchase a book on the latter topic.

Acknowledgments. We are grateful to Alina Arbitman and Ron Rothblum for their comments and suggestions regarding this primer.

Oded Goldreich
Weizmann Institute of Science

Chapter 1

Introduction

The "question of randomness" has been puzzling thinkers for ages. Aspects of this question range from philosophical doubts regarding the existence of randomness (in the world) and reflections on the meaning of randomness (in our thinking) to technical questions regarding the measuring of randomness. Among many other things, the second half of the twentieth century has witnessed the development of three theories of randomness, which address different aspects of the foregoing question.

The first theory (cf., [16]), initiated by Shannon [63], views randomness as representing *uncertainty*, which in turn is modeled by a probability distribution on the possible values of the missing data. Indeed, Shannon's Information Theory is rooted in probability theory. Information Theory focuses on distributions that are not perfectly random (i.e., encode information in a redundant manner), and characterizes perfect randomness as the extreme case in which the uncertainty is maximized (i.e., in this case there is no redundancy at all). Thus, perfect randomness is associated with a unique distribution– the uniform one. In particular, by definition, one cannot (deterministically) generate such perfect random strings from shorter random seeds.

The second theory (cf., [41, 42]), initiated by Solomonoff [64], Kolmogorov [38], and Chaitin [14], views randomness as representing the lack of structure, which in turn is reflected in the length of the most succinct (effective) description of the object. The notion of a succinct and *effective description* refers to a process that transforms the succinct description to an explicit one. Indeed, this theory of randomness is rooted in computability theory and specifically in the notion of a universal language (equiv., universal machine or computing device). It measures the randomness (or complexity) of objects in terms of the shortest program (for a fixed universal machine) that generates the object.[1] Like Shannon's theory, Kolmogorov Complexity is quantitative and perfect random objects appear as an extreme case. However, following Kolmogorov's approach one may say that a single object, rather than a distribution over objects, is perfectly random. Still, by definition, one cannot (deterministically) generate strings of high Kolmogorov Complexity from short random seeds.

[1] We mention that Kolmogorov's approach is inherently intractable (i.e., Kolmogorov Complexity is uncomputable).

1.1 The Third Theory of Randomness

The third theory, which is the focus of the current primer, views randomness as an effect on an observer and thus as being relative to the *observer's abilities* (of analysis). The observer's abilities are captured by its computational abilities (i.e., the complexity of the processes that the observer may apply), and hence this theory of randomness is rooted in complexity theory. This theory of randomness is explicitly aimed at providing a notion of randomness that, unlike the previous two notions, allows for an efficient (and deterministic) generation of random strings from shorter random seeds. The heart of this theory is the suggestion to view objects as equal if they cannot be distinguished by any efficient procedure. Consequently, a distribution that cannot be efficiently distinguished from the uniform distribution will be considered random (or rather called pseudorandom). Thus, randomness is not an "inherent" property of objects (or distributions) but is rather relative to an observer (and its computational abilities). To illustrate this perspective, let us consider the following mental experiment.

> Alice and Bob play "heads or tails" in one of the following four ways. In each of them, Alice flips an unbiased coin and Bob is asked to guess its outcome *before* the coin hits the floor. The alternative ways differ by the knowledge Bob has before making his guess.
>
> In the first alternative, Bob has to announce his guess before Alice flips the coin. Clearly, in this case Bob wins with probability 1/2.
>
> In the second alternative, Bob has to announce his guess while the coin is spinning in the air. Although the outcome is *determined in principle* by the motion of the coin, Bob does not have accurate information on the motion. Thus we believe that, also in this case, Bob wins with probability 1/2.
>
> The third alternative is similar to the second, except that Bob has at his disposal sophisticated equipment capable of providing accurate *information* on the coin's motion as well as on the environment effecting the outcome. However, Bob cannot process this information in time to improve his guess.
>
> In the fourth alternative, Bob's recording equipment is directly connected to a *powerful computer* programmed to solve the motion equations and output a prediction. It is conceivable that in such a case Bob can substantially improve his guess of the outcome of the coin.

We conclude that the randomness of an event is relative to the information and computing resources at our disposal. At the extreme, even events that are fully determined by public information may be perceived as random events by an observer who lacks the relevant information and/or the ability to process it. Our focus will be on the lack of sufficient processing power, and not on the lack of sufficient information. The lack of sufficient processing power may be due either to the formidable amount of computation required (for analyzing the event in question) or to the fact that the observer happens to be very limited.

A natural notion of pseudorandomness arises: a distribution is *pseudorandom* if no efficient procedure can distinguish it from the uniform distribution, where efficient procedures are associated with (probabilistic) polynomial-time algorithms. This specific notion of pseudorandomness is indeed the most fundamental one, and much of this text is focused on it. Weaker notions of pseudorandomness arise as well – they refer to indistinguishability by weaker procedures such as space-bounded algorithms, constant-depth circuits, etc. Stretching this approach even further one may consider algorithms that are designed (on purpose so) not to distinguish even weaker forms of "pseudorandom" sequences from random ones. Such algorithms arise naturally when trying to convert some natural randomized algorithm into deterministic ones; see Chapter 5.

The preceding discussion has focused on one aspect of the pseudorandomness question – the resources or type of the observer (or potential distinguisher). Another important aspect is whether such pseudorandom sequences can be generated from much shorter ones, and at what cost (or complexity). A natural approach requires the generation process to be efficient, and furthermore to be fixed before the specific observer is determined. Coupled with the aforementioned strong notion of pseudorandomness, this yields the archetypical notion of pseudorandom generators – those operating in (fixed) polynomial-time and producing sequences that are indistinguishable from uniform ones by *any* polynomial-time observer. In particular, this means that the distinguisher is allowed more resources than the generator. Such (general-purpose) pseudorandom generators (discussed in Chapter 2) allow one to decrease the randomness complexity of *any efficient application*, and are thus of great relevance to randomized algorithms and cryptography. The term *general-purpose* is meant to emphasize the fact that the same generator is good for all efficient applications, including those that consume more resources than the generator itself.

Although general-purpose pseudorandom generators are very appealing, there are important reasons for considering also the opposite relation between the complexities of the generation and distinguishing tasks; that is, allowing the pseudorandom generator to use more resources (e.g., time or space) than the observer it tries to fool. This alternative is natural in the context of derandomization (i.e., converting randomized algorithms to deterministic ones), where the crucial step is replacing the random input of an algorithm by a pseudorandom input, which in turn can be generated based on a much shorter random seed. In particular, when derandomizing a probabilistic polynomial-time algorithm, the observer (to be fooled by the generator) is a fixed algorithm. In this case employing a more complex generator merely means that the complexity of the derived deterministic algorithm is dominated by the complexity of the generator (rather than by the complexity of the original randomized algorithm). Needless to say, allowing the generator to use more resources than the observer that it tries to fool makes the task of designing pseudorandom generators potentially easier, and enables derandomization results that are not known when using general-purpose pseudorandom generators. The usefulness of this approach is demonstrated in Chapters 3 through 5.

We note that the goal of all types of pseudorandom generators is to allow the generation of "sufficiently random" sequences based on much shorter random seeds. Thus, pseudorandom generators offer significant savings in the randomness complexity of various applications (and in some cases eliminating randomness altogether). Saving on randomness is valuable because many applications are severely limited in their ability to generate or obtain truly random bits. Furthermore, typically, generating truly random bits is significantly more expensive than standard computation steps. Thus, randomness is a computational resource that should be considered on top of time complexity (analogously to the consideration of space complexity).

1.2 Organization of the Primer

We start by presenting some standard conventions (see Section 1.3). Next, in Section 1.4, we present the general paradigm underlying the various notions of pseudorandom generators. The archetypical case of general-purpose pseudorandom generators is presented in Chapter 2. We then turn to alternative notions of pseudorandom generators: generators that suffice for the derandomization of complexity classes such as $\mathcal{BPP}$ are discussed in Chapter 3; pseudorandom generators in the domain of space-bounded computations are discussed in Chapter 4; and several notions of special-purpose generators are discussed in Chapter 5.

The text is organized to facilitate the possibility of focusing on the notion of general-purpose pseudorandom generators (presented in Chapter 2). This notion is most relevant to computer science at large, and consequently it is most relevant to other sciences. Furthermore, the technical details presented in Chapter 2 are relatively simpler than those presented in Chapters 3 and 4.

The appendices. For the benefit of readers who are less familiar with computer science, we augment the foregoing material with six appendices. Appendix A provides a basic treatment of hashing functions, which are used in Section 4.2 and are related to the limited-independence generators discussed in Section 5.1. Appendix B provides a brief introduction to the notion of randomness extractors, which are of natural interest as well as being used in Section 4.2. Appendix C provides a proof of a key result that is closely related to the material of Section 2.5. Appendix D provides three illustrations to the use of randomness in computation. Appendix E presents a couple of basic cryptographic applications of pseudorandom functions, which are treated in Section 2.7.2. Appendix F provides definitions of some basic complexity classes.

Relation to complexity theory. The study of pseudorandom generators is part of complexity theory, and the interested reader is encouraged to further explore the connections between pseudorandomness and complexity theory at large (cf. e.g., [24]). In fact, the current primer is an abbreviated (and revised) version of [24, Chap. 8].

Preliminaries. We assume a basic familiarity with computational complexity; that is, we assume that the reader is comfortable with the notion of efficient algorithms and their association with polynomial-time algorithms (see, e.g., [24]). We also assume that the reader is aware that very basic questions about the nature of efficient computation are wide open (e.g., most notably, the P-vs-NP Question).

We also assume a basic familiarity with elementary probability theory (see any standard textbook or brief reviews in [46, 47, 24]) and randomized algorithms (see, e.g., either [47, 46] or [24, Chap. 6]). In particular, standard conventions regarding random variables (presented next) will be extensively used.

1.3 Standard Conventions

Throughout the entire text we refer only to *discrete* probability distributions. Specifically, the underlying probability space consists of the set of all strings of a certain length ℓ, taken with uniform probability distribution. That is, the sample space is the set of all ℓ-bit long strings, and each such string is assigned probability measure $2^{-\ell}$. Traditionally, *random variables* are defined as functions from the sample space to the reals. Abusing the traditional terminology, we use the term random variable also when referring to functions mapping the sample space into the set of binary strings. We often do not specify the probability space, but rather talk directly about random variables. For example, we may say that X is a random variable assigned values in the set of all strings such that $\Pr[X=00] = \frac{1}{4}$ and $\Pr[X=111] = \frac{3}{4}$. (Such a random variable may be defined over the sample space $\{0,1\}^2$ such that $X(11) = 00$ and $X(00) = X(01) = X(10) = 111$.) One important case of a random variable is the output of a randomized process (e.g., a probabilistic polynomial-time algorithm).

All of our probabilistic statements refer to random variables that are defined beforehand. Typically, we may write $\Pr[f(X) = 1]$, where X is a random variable defined beforehand (and f is a function). An important convention is that *all occurrences of the same symbol in a probabilistic statement refer to the same* (unique) *random variable.* Hence, if $B(\cdot,\cdot)$ is a Boolean expression depending on two variables, and X is a random variable, then $\Pr[B(X,X)]$ denotes the probability that $B(x,x)$ holds when x is chosen with probability $\Pr[X=x]$. For example, for every random variable X, we have $\Pr[X=X] = 1$. We stress that if we wish to discuss the probability that $B(x,y)$ holds when x and y are chosen independently with identical probability distribution, then we will define *two* independent random variables each with the same probability distribution. Hence, if X and Y are two independent random variables, then $\Pr[B(X,Y)]$ denotes the probability that $B(x,y)$ holds when the pair (x,y) is chosen with probability $\Pr[X=x] \cdot \Pr[Y=y]$. For example, for every two independent random variables, X and Y, we have $\Pr[X=Y] = 1$ only if both X and Y are trivial (i.e., assign the entire probability mass to a single string).

Throughout the entire text, U_n denotes a random variable uniformly distributed over the set of all strings of length n. Namely, $\Pr[U_n=\alpha]$ equals 2^{-n} if $\alpha \in \{0,1\}^n$ and equals 0 otherwise. We often refer to the distribution of U_n as the uniform distribution (neglecting to qualify that it is uniform over $\{0,1\}^n$). In addition, we occasionally use random variables (arbitrarily) distributed over $\{0,1\}^n$ or $\{0,1\}^{\ell(n)}$, for some function $\ell:\mathbb{N}\to\mathbb{N}$. Such random variables are typically denoted by X_n, Y_n, Z_n, etc. We stress that in some cases X_n is distributed over $\{0,1\}^n$, whereas in other cases it is distributed over $\{0,1\}^{\ell(n)}$, for some function ℓ (which is typically a polynomial). We often talk about probability ensembles, which are infinite sequences of random variables $\{X_n\}_{n\in\mathbb{N}}$ such that each X_n ranges over strings of length bounded by a polynomial in n.

Statistical difference. The statistical distance (a.k.a variation distance) between the random variables X and Y is defined as

$$\frac{1}{2} \cdot \sum_v |\Pr[X = v] - \Pr[Y = v]| \;=\; \max_S\{\Pr[X \in S] - \Pr[Y \in S]\} \qquad (1.1)$$

(see Exercise 1.1). We say that X is δ-close (resp., δ-far) to Y if the statistical distance between them is at most (resp., at least) δ.

1.4 The General Paradigm

We advocate a unified view of various notions of pseudorandom generators. That is, we view these notions as incarnations of a general abstract paradigm, to be presented in this section. A reader who is interested only in one of these incarnations may still use this section as a general motivation towards the specific definitions used later. On the other hand, some readers may prefer reading this section after studying one of the specific incarnations.

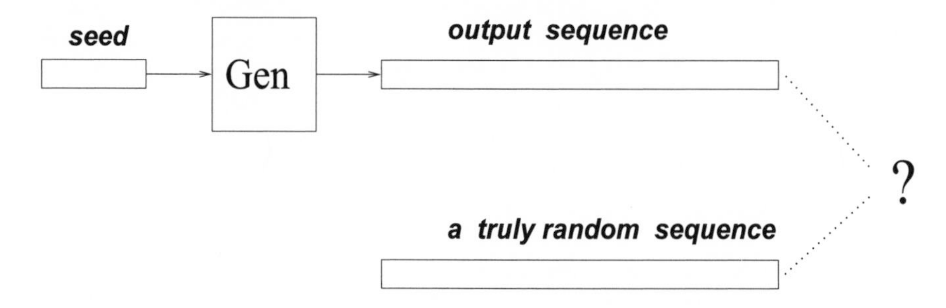

Figure 1.1: Pseudorandom generators – an illustration.

1.4.1 Three fundamental aspects

A generic formulation of pseudorandom generators consists of specifying three fundamental aspects – the *stretch measure* of the generators; the class of distinguishers that the generators are supposed to fool (i.e., the algorithms with respect to which the *computational indistinguishability* requirement should hold); and the resources that the generators are allowed to use (i.e., their own *computational complexity*). Let us elaborate.

Stretch function: A necessary requirement from any notion of a pseudorandom generator is that the generator is a *deterministic algorithm* that stretches short strings, called seeds, into longer output sequences.[2] Specifically, this algorithm stretches k-bit long seeds into $\ell(k)$-bit long outputs, where $\ell(k) > k$. The function $\ell:\mathbb{N}\rightarrow\mathbb{N}$ is called the stretch measure (or stretch function) of the generator. In some settings the specific stretch measure is immaterial (e.g., see Section 2.4).

Computational Indistinguishability: A necessary requirement from any notion of a pseudorandom generator is that the generator "fools" some non-trivial algorithms. That is, it is required that any algorithm taken from a predetermined class

[2]Indeed, the seed represents the randomness that is used in the generation of the output sequences; that is, the randomized generation process is decoupled into a deterministic algorithm and a random seed. This decoupling facilitates the study of such processes.

of interest cannot distinguish the output produced by the generator (when the generator is fed with a uniformly chosen seed) from a uniformly chosen sequence. Thus, we consider a class $\mathcal{D}$ of distinguishers (e.g., probabilistic polynomial-time algorithms) and a class $\mathcal{F}$ of (threshold) functions (e.g., reciprocals of positive polynomials), and require that the generator G satisfies the following: For any $D \in \mathcal{D}$, any $f \in \mathcal{F}$, and for all sufficiently large k it holds that

$$|\Pr[D(G(U_k)) = 1] - \Pr[D(U_{\ell(k)}) = 1]| < f(k), \quad (1.2)$$

where U_n denotes the uniform distribution over $\{0,1\}^n$, and the probability is taken over U_k (resp., $U_{\ell(k)}$) as well as over the coin tosses of algorithm D in case it is probabilistic. The reader may think of such a distinguisher, D, as an observer who tries to tell whether the "tested string" is a random output of the generator (i.e., distributed as $G(U_k)$) or is a truly random string (i.e., distributed as $U_{\ell(k)}$). The condition in Eq. (1.2) requires that D cannot make a meaningful decision; that is, ignoring a negligible difference (represented by $f(k)$), D's verdict is the same in both cases.[3] The archetypical choice is that $\mathcal{D}$ is the set of all probabilistic polynomial-time algorithms, and $\mathcal{F}$ is the set of all functions that are the reciprocal of some positive polynomial.

We note that there is a clear tension between the stretching and the computational indistinguishability conditions. Indeed, as shown in Exercise 1.2, the output of any pseudorandom generator is "statistically distinguishable" from the corresponding uniform distribution. However, there is hope that a restricted class of (computationally bounded) distinguishers cannot detect the (statistical) difference; that is, be fooled by some suitable generators. In fact, placing no computational requirements on the generator (or, alternatively, imposing very mild requirements such as upper-bounding the running-time by a double-exponential function), yields "generators" that can fool any subexponential-size circuit family (see Exercise 1.3). However, we are interested in the complexity of the generation process, which is the aspect addressed next.

Complexity of Generation: This aspect refers to the complexity of the generator itself, when viewed as an algorithm. That is, here we refer to the resources used by the generator (e.g., its time and/or space complexity). The archetypical choice is that the generator has to work in polynomial-time (i.e., make a number of steps that is polynomial in the length of its input – the seed). Other choices will be discussed as well.

1.4.2 Notational conventions

We will consistently use k for denoting the length of the seed of a pseudorandom generator, and $\ell(k)$ for denoting the length of the corresponding output. In some cases, this makes our presentation a little more cumbersome, where in these cases

[3]The class of threshold functions $\mathcal{F}$ should be viewed as determining the class of noticeable probabilities (as a function of k). Thus, we require certain functions (i.e., those presented on the l.h.s of Eq. (1.2)) to be smaller than any noticeable function *on all but finitely many integers*. We call the former functions negligible. Note that a function may be neither noticeable nor negligible (e.g., it may be smaller than any noticeable function on infinitely many values and yet larger than some noticeable function on infinitely many other values).

it is more natural to focus on a different parameter (e.g., the length of the pseudorandom sequence) and let the seed-length be a function of the latter. However, our choice has the advantage of focusing attention on the fundamental parameter of pseudorandom generation process – the length of the random seed. We note that whenever a pseudorandom generator is used to "derandomize" an algorithm, n will denote the length of the input to this algorithm, and k will be selected as a function of n.

1.4.3 Some instantiations of the general paradigm

Two important instantiations of the notion of pseudorandom generators relate to polynomial-time distinguishers.

1. General-purpose pseudorandom generators correspond to the case where the generator itself runs in polynomial-time and needs to withstand *any probabilistic polynomial-time distinguisher*, including distinguishers that run for more time than the generator. Thus, the same generator may be used safely in any efficient application. (This notion is treated in Chapter 2.)

2. In contrast, pseudorandom generators intended for derandomization may run for more time than the distinguisher, which is viewed as a fixed circuit having size that is upper-bounded by a fixed polynomial. (This notion is treated in Chapter 3.)

In addition, the general paradigm may be instantiated by focusing on the space-complexity of the potential distinguishers (and the generator), rather than on their time-complexity. Furthermore, one may also consider distinguishers that merely reflect probabilistic properties such as pairwise independence, small-bias, and hitting frequency.

Notes

Our presentation, which views vastly different notions of pseudorandom generators as incarnations of a general paradigm, has emerged mostly in retrospect. We note that, while the historical study of the various notions was mostly unrelated at a technical level, the case of general-purpose pseudorandom generators served as a source of inspiration to most of the other cases. In particular, the concept of computational indistinguishability, the connection between hardness and pseudorandomness, and the equivalence between pseudorandomness and unpredictability, appeared first in the context of general-purpose pseudorandom generators (and inspired the development of "generators for derandomization" and "generators for space bounded machines"). Indeed, the study of the special-purpose generators (see Chapter 5) was unrelated to all of these.

We mention that an alternative treatment of pseudorandomness, which puts more emphasis on the relation between various techniques, is provided in [68]. In particular, the latter text highlights the connections between information theoretic and computational phenomena (e.g., randomness extractors and canonical derandomizers), while the current text tends to decouple the two.

Exercises

Exercise 1.1 Prove the equality in Eq. (1.1).

Guideline: Let S be the set of strings having a larger probability under the first distribution.

Exercise 1.2 Show that the output of any pseudorandom generator is "statistically distinguishable" from the corresponding uniform distribution; that is, show that, for any stretch function ℓ and any generator G of stretch ℓ, the statistical difference between $G(U_k)$ and $U_{\ell(k)}$ is at least $1 - 2^{-(\ell(k)-k)}$.

Exercise 1.3 Show that placing no computational requirements on the generator enables unconditional results regarding "generators" that fool any family of subexponential-size circuits. That is, making no computational assumptions, prove that there exist functions $G : \{0,1\}^* \to \{0,1\}^*$ such that $\{G(U_k)\}_{k \in \mathbb{N}}$ is (strongly) pseudorandom, while $|G(s)| = 2|s|$ for every $s \in \{0,1\}^*$. Furthermore, show that G can be computed in double-exponential time.

Guideline: Use the Probabilistic Method (cf. [6]). First, for any fixed circuit $C : \{0,1\}^n \to \{0,1\}$, upper-bound the probability that for a random set $S \subset \{0,1\}^n$ of size $2^{n/2}$ the absolute value of $\Pr[C(U_n) = 1] - (|\{x \in S : C(x) = 1\}|/|S|)$ is larger than $2^{-n/8}$. Next, using a union bound, prove the existence of a set $S \subset \{0,1\}^n$ of size $2^{n/2}$ such that no circuit of size $2^{n/5}$ can distinguish a uniformly distributed element of S from a uniformly distributed element of $\{0,1\}^n$, where distinguishing means with a probability gap of at least $2^{-n/8}$.

Chapter 2

General-Purpose Pseudorandom Generators

Randomness is playing an increasingly important role in computation: It is frequently used in the design of sequential, parallel and distributed algorithms, and it is of course central to cryptography. Whereas it is convenient to design such algorithms making free use of randomness, it is also desirable to minimize the usage of randomness in real implementations. Thus, general-purpose pseudorandom generators (as defined next) are a key ingredient in an "algorithmic tool-box" – they provide an automatic compiler of programs written with free usage of randomness into programs that make an economical use of randomness.

Organization of this chapter. Since this is a relatively long chapter, a short roadmap seems appropriate. In Section 2.1 we provide the basic definition of general-purpose pseudorandom generators, and in Section 2.2 we describe their archetypical application (which was alluded to in the former paragraph). In Section 2.3 we provide a wider perspective on the notion of computational indistinguishability that underlies the basic definition, and in Section 2.4 we justify the little concern (shown in Section 2.1) regarding the specific stretch function. In Section 2.5 we address the existence of general-purpose pseudorandom generators. In Section 2.6 we motivate and discuss a non-uniform version of computational indistinguishability. We conclude by reviewing other variants and reflecting on various conceptual aspects of the notions discussed in this chapter (see Sections 2.7 and 2.8, resp.).

2.1 The Basic Definition

Loosely speaking, general-purpose pseudorandom generators are efficient deterministic programs that expand short randomly selected seeds into longer pseudorandom bit sequences, where the latter are defined as computationally indistinguishable from truly random sequences by *any* efficient algorithm. Identifying efficiency with polynomial-time operation, this means that the generator (being a fixed algorithm) works within *some fixed* polynomial-time, whereas the distinguisher may be *any* algorithm that runs in polynomial-time. Thus, the distinguisher is potentially more complex than the generator; for example, the distinguisher *may* run in time that is cubic in the running-time of the generator. Furthermore, to facilitate the development of this theory, we allow the distinguisher to be probabilistic (whereas the generator remains deterministic as stated previously). We require that such distinguishers cannot tell the output of the generator from a truly random string of similar length, or rather that the difference that such distinguishers may detect (or "sense") is negligible. Here a negligible function is a function that vanishes faster than the reciprocal of any positive polynomial.[1]

Definition 2.1 (general-purpose pseudorandom generator): *A deterministic polynomial-time algorithm G is called a* pseudorandom generator *if there exists a* stretch function, *$\ell : \mathbb{N}\to\mathbb{N}$ (satisfying $\ell(k) > k$ for all k), such that for any probabilistic polynomial-time algorithm D, for any positive polynomial p, and for all sufficiently large k it holds that*

$$|\,\mathsf{Pr}[D(G(U_k)) = 1] - \mathsf{Pr}[D(U_{\ell(k)}) = 1]\,| < \frac{1}{p(k)} \tag{2.1}$$

where U_n denotes the uniform distribution over $\{0,1\}^n$ and the probability is taken over U_k (resp., $U_{\ell(k)}$) as well as over the internal coin tosses of D.

Thus, Definition 2.1 is derived from the generic framework (presented in Section 1.4) by taking the class of distinguishers to be the set of all probabilistic polynomial-time algorithms, and taking the class of (noticeable) threshold functions to be the set of all functions that are the reciprocals of some positive polynomial.[2] Indeed, the principles underlying Definition 2.1 were discussed in Section 1.4 (and will be further discussed in Section 2.3).

We note that Definition 2.1 does not make any requirement regarding the stretch function $\ell:\mathbb{N}\to\mathbb{N}$, except for the generic requirement that $\ell(k) > k$ for all k. Needless to say, the larger ℓ is, the more useful the pseudorandom generator is. Of course, ℓ is upper-bounded by the running-time of the generator (and hence by a polynomial). In Section 2.4 we show that any pseudorandom generator (even one having minimal stretch $\ell(k) = k + 1$) can be used for constructing a pseudorandom generator having any desired (polynomial) stretch function. But before doing so, we rigorously discuss the "saving in randomness" offered by pseudorandom generators, and provide a wider perspective on the notion of computational indistinguishability that underlies Definition 2.1.

2.2 The Archetypical Application

We note that "pseudorandom number generators" appeared with the first computers, and have been used ever since for generating random choices (or samples) for

---

[1]Definition 2.1 requires that the functions representing the distinguishing gap of certain algorithms should be smaller than the reciprocal of any positive polynomial for all but finitely many k's, and the former functions are called *negligible*. The notion of negligible probability is robust in the sense that any event that occurs with negligible probability will occur with negligible probability also when the experiment is repeated a "feasible" (i.e., polynomial) number of times.

[2]The latter choice is naturally coupled with the association of efficient computation with polynomial-time algorithms: An event that occurs with noticeable probability occurs almost always when the experiment is repeated a "feasible" (i.e., polynomial) number of times.

various applications. However, typical implementations use generators that are not pseudorandom according to Definition 2.1. Instead, at best, these generators are shown to pass *some* ad-hoc statistical test (cf., [37]). We warn that the fact that a "pseudorandom number generator" passes some statistical tests, does not mean that it will pass a new test and that it will be good for a future (untested) application. Needless to say, the approach of subjecting the generator to some ad-hoc tests fails to provide general results of the form "for *all* practical purposes using the output of the generator is as good as using truly unbiased coin tosses." In contrast, the approach encompassed in Definition 2.1 aims at such generality, and in fact is tailored to obtain it: The notion of computational indistinguishability, which underlines Definition 2.1, covers all possible efficient applications and guarantees that for all of them pseudorandom sequences are as good as truly random ones. Indeed, any efficient randomized algorithm maintains its performance when its internal coin tosses are substituted by a sequence generated by a pseudorandom generator. This substitution is spelled out next.

Construction 2.2 (typical application of pseudorandom generators): *Let G be a pseudorandom generator with stretch function $\ell : \mathbb{N} \to \mathbb{N}$. Let A be a probabilistic polynomial-time algorithm, and let $\rho : \mathbb{N} \to \mathbb{N}$ denote its randomness complexity. Denote by $A(x,r)$ the output of A on input x and the coin toss sequence $r \in \{0,1\}^{\rho(|x|)}$. Consider the following randomized algorithm, denoted A_G:*

> *On input x, set $k = k(|x|)$ to be the smallest integer such that $\ell(k) \geq \rho(|x|)$, uniformly select $s \in \{0,1\}^k$, and output $A(x,r)$, where r is the $\rho(|x|)$-bit long prefix of $G(s)$.*

That is, $A_G(x,s) = A(x,G'(s))$, for $|s| = k(|x|) = \mathrm{argmin}_i\{\ell(i) \geq \rho(|x|)\}$, where $G'(s)$ is the $\rho(|x|)$-bit long prefix of $G(s)$.

Thus, using A_G instead of A, the randomness complexity is reduced from ρ to $\ell^{-1} \circ \rho$, while (as we show next) it is infeasible to find inputs (i.e., x's) on which the *noticeable behavior* of A_G is different from that of A. For example, if $\ell(k) = k^2$, then the randomness complexity is reduced from ρ to $\sqrt{\rho}$. We stress that the pseudorandom generator G is *universal*; that is, it can be applied to reduce the randomness complexity of *any* probabilistic polynomial-time algorithm A. The following proposition asserts that it is infeasible to find an input on which A_G behaves differently than A.

Proposition 2.3 (analysis of Construction 2.2): *Let A, ρ and G be as in Construction 2.2, and suppose that $\rho : \mathbb{N} \to \mathbb{N}$ is one-to-one. Then, for every pair of probabilistic polynomial-time algorithms,* a finder *F and* a tester *T, every positive polynomial p and all sufficiently long n, it holds that*

$$\sum_{x \in \{0,1\}^n} \mathsf{Pr}[F(1^n) = x] \cdot |\,\Delta_{A,T}(x)\,| \;<\; \frac{1}{p(n)} \tag{2.2}$$

where $\Delta_{A,T}(x) \stackrel{\text{def}}{=} \mathsf{Pr}[T(x, A(x, U_{\rho(|x|)})) = 1] - \mathsf{Pr}[T(x, A_G(x, U_{k(|x|)})) = 1]$, and the probabilities are taken over the U_m's as well as over the internal coin tosses of the algorithms F and T.

Algorithm F represents a potential attempt to find an input x on which the output of A_G is distinguishable from the output of A, where F is given a length parameter n (in unary) and is required to produce a corresponding n-bit string in poly(n)-time.[3] This "attempt" (represented by F) may be benign (as in the case that a user employs algorithm A_G on inputs that are generated by some probabilistic polynomial-time application), but it may also be adversarial (as in the case that a user employs algorithm A_G on inputs that are provided by a potentially malicious party). The potential tester, denoted T, represents the potential use of the output of algorithm A_G, and captures the requirement that this output be as good as a corresponding output produced by A. Thus, T is given x as well as the corresponding output produced either by $A_G(x) \stackrel{\text{def}}{=} A(x, U_{k(|x|)})$ or by $A(x) = A(x, U_{\rho(|x|)})$, and it is required that T cannot tell the difference. In the case that A is a probabilistic polynomial-time *decision procedure*, this means that it is infeasible to find an x on which A_G decides incorrectly (i.e., differently than A). In the case that A is a *search procedure for some NP-relation*, it is infeasible to find an x on which A_G outputs a wrong solution. For details, see Exercise 2.1.

Proof Sketch: The proposition is proven by showing that any triple (A, F, T) violating the claim can be converted into an algorithm D that distinguishes the output of G from the uniform distribution, in contradiction to the hypothesis. The key observation is that for every $x \in \{0,1\}^n$ it holds that

$$\Delta_{A,T}(x) = \mathsf{Pr}[T(x, A(x, U_{\rho(n)}))=1] - \mathsf{Pr}[T(x, A(x, G'(U_{k(n)})))=1], \tag{2.3}$$

where $G'(s)$ is the $\rho(n)$-bit long prefix of $G(s)$. Thus, a method for finding a string x such that $|\Delta_{A,T}(x)|$ is large, yields a way of distinguishing $U_{\ell(k(n))}$ from $G(U_{k(n)})$; that is, given a sample $r \in \{0,1\}^{\ell(k(n))}$ and using such a string $x \in \{0,1\}^n$, the distinguisher outputs $T(x, A(x, r'))$, where r' is the $\rho(n)$-bit long prefix of r. Indeed, we shall show that the violation of Eq. (2.2), which refers to $\mathsf{E}_{x \leftarrow F(1^n)}[|\Delta_{A,T}(x)|]$, yields a violation of the hypothesis that G is a pseudorandom generator (by finding an adequate string x and using it). This intuitive argument requires a slightly careful implementation, which is provided next.

As a warm-up, consider the following algorithm D. On input r (taken from either $U_{\ell(k(n))}$ or $G(U_{k(n)})$), algorithm D first obtains $x \leftarrow F(1^n)$, where n can be obtained easily from $|r|$ (because ρ is one-to-one and $1^n \mapsto \rho(n)$ is computable via A). Next, D obtains $y = A(x, r')$, where r' is the $\rho(|x|)$-bit long prefix of r. Finally, D outputs $T(x, y)$. Note that D is implementable in probabilistic polynomial-time, and that

$$\begin{aligned} D(U_{\ell(k(n))}) &\equiv T(X_n, A(X_n, U_{\rho(n)}))\,, \text{ where } X_n \stackrel{\text{def}}{=} F(1^n), \\ D(G(U_{k(n)})) &\equiv T(X_n, A(X_n, G'(U_{k(n)})))\,, \text{ where } X_n \stackrel{\text{def}}{=} F(1^n). \end{aligned}$$

Using Eq. (2.3), it follows that $\mathsf{Pr}[D(U_{\ell(k(n))}) = 1] - \mathsf{Pr}[D(G(U_{k(n)})) = 1]$ equals $\mathsf{E}[\Delta_{A,T}(F(1^n))]$, which implies that $\mathsf{E}[\Delta_{A,T}(F(1^n))]$ must be negligible (because otherwise we derive a contradiction to the hypothesis that G is a pseudorandom generator). This yields a weaker version of the proposition asserting that $\mathsf{E}[\Delta_{A,T}(F(1^n))]$ is negligible (rather than that $\mathsf{E}[|\Delta_{A,T}(F(1^n))|]$ is negligible).

[3]Indeed, providing n in unary (i.e., as 1^n) and postulating that F runs in polynomial-time implies that F should find $x \in \{0,1\}^n$ in poly(n)-time.

In order to prove that $\mathsf{E}[|\Delta_{A,T}(F(1^n))|]$ (rather than $\mathsf{E}[\Delta_{A,T}(F(1^n))]$) is negligible, we need to modify D a little. Note that the source of trouble is that $\Delta_{A,T}(\cdot)$ may be positive on some x's and negative on others, and thus it may be the case that $\mathsf{E}[\Delta_{A,T}(F(1^n))]$ is small (due to cancelations) even if $\mathsf{E}[|\Delta_{A,T}(F(1^n))|]$ is large. This difficulty can be overcome by determining the sign of $\Delta_{A,T}(\cdot)$ on $x = F(1^n)$ and changing the outcome of D accordingly; that is, the modified D will output $T(x, A(x, r'))$ if $\Delta_{A,T}(x) > 0$ and $1 - T(x, A(x, r'))$ otherwise. Thus, in each case, the contribution of x to the distinguishing gap of the modified D will be $|\Delta_{A,T}(x)|$. We further note that if $|\Delta_{A,T}(x)|$ is small, then it does not matter much whether we act as in the case of $\Delta_{A,T}(x) > 0$ or in the case of $\Delta_{A,T}(x) \leq 0$. Thus, it suffices to correctly determine the sign of $\Delta_{A,T}(x)$ in the case that $|\Delta_{A,T}(x)|$ is large, which is certainly a feasible (approximation) task. Details can be found in [24, Sec. 8.2.2]. □

Conclusion. Although Proposition 2.3 refers to standard probabilistic polynomial-time algorithms, a similar construction and analysis applied to any efficient randomized process (i.e., any efficient multi-party computation). Any such process preserves its behavior when replacing its perfect source of randomness (postulated in its analysis) by a pseudorandom sequence (which may be used in the implementation). Thus, given a pseudorandom generator with a large stretch function, *one can considerably reduce the randomness complexity of any efficient application.*

2.3 Computational Indistinguishability

In this section we spell out (and study) the definition of computational indistinguishability that underlies Definition 2.1.

2.3.1 The general formulation

The (general formulation of the) definition of computational indistinguishability refers to *arbitrary* probability ensembles. Here a probability ensemble is an infinite sequence of random variables $\{Z_n\}_{n\in\mathbb{N}}$ such that each Z_n ranges over strings of length that is polynomially related to n (i.e., there exists a polynomial p such that for every n it holds that $|Z_n| \leq p(n)$ and $p(|Z_n|) \geq n$). We say that $\{X_n\}_{n\in\mathbb{N}}$ and $\{Y_n\}_{n\in\mathbb{N}}$ are computationally indistinguishable if for every feasible algorithm A the difference $d_A(n) \stackrel{\text{def}}{=} |\mathsf{Pr}[A(X_n)=1] - \mathsf{Pr}[A(Y_n)=1]|$ is a negligible function in n. That is:

Definition 2.4 (computational indistinguishability): *The probability ensembles $\{X_n\}_{n\in\mathbb{N}}$ and $\{Y_n\}_{n\in\mathbb{N}}$ are* computationally indistinguishable *if for every probabilistic polynomial-time algorithm D, every positive polynomial p, and all sufficiently large n, it holds that*

$$|\mathsf{Pr}[D(X_n)=1] - \mathsf{Pr}[D(Y_n)=1]| < \frac{1}{p(n)} \tag{2.4}$$

where the probabilities are taken over the relevant distribution (i.e., either X_n or Y_n) *and over the internal coin tosses of algorithm D. The l.h.s. of Eq. (2.4), when viewed as a function of n, is often called the* distinguishing gap of D, *where $\{X_n\}_{n\in\mathbb{N}}$ and $\{Y_n\}_{n\in\mathbb{N}}$ are understood from the context.*

We can think of D as representing somebody who wishes to distinguish two distributions (based on a given sample drawn from one of the distributions), and think of the output "1" as representing D's verdict that the sample was drawn according to the first distribution. Saying that the two distributions are computationally indistinguishable means that if D is a feasible procedure, then its verdict is not really meaningful (because the verdict is almost as often 1 when the sample is drawn from the first distribution as when the sample is drawn from the second distribution). We comment that the absolute value in Eq. (2.4) can be omitted without affecting the definition (see Exercise 2.2), and we will often do so without warning.

In Definition 2.1, we required that the probability ensembles $\{G(U_k)\}_{k\in\mathbb{N}}$ and $\{U_{\ell(k)}\}_{k\in\mathbb{N}}$ be computationally indistinguishable. Indeed, an important special case of Definition 2.4 is when one ensemble is uniform, and in such a case we call the other ensemble **pseudorandom**.

2.3.2 Relation to statistical closeness

Two probability ensembles, $\{X_n\}_{n\in\mathbb{N}}$ and $\{Y_n\}_{n\in\mathbb{N}}$, are said to be **statistically close** (or statistically indistinguishable) if for every positive polynomial p and all sufficiently large n the variation distance between X_n and Y_n (i.e., $\frac{1}{2}\sum_z |\mathsf{Pr}[X_n=z]-\mathsf{Pr}[Y_n=z]|$) is bounded above by $1/p(n)$. Clearly, any two probability ensembles that are statistically close are computationally indistinguishable. Needless to say, this is a trivial case of computational indistinguishability, which is due to information theoretic reasons. In contrast, we shall be interested in *non-trivial cases* (of computational indistinguishability), which correspond to probability ensembles that are statistically far apart.

Indeed, as claimed in Section 1.4 (see Exercise 1.3), there exist probability ensembles that are statistically far apart and yet are computationally indistinguishable. However, at least one of the two probability ensembles in Exercise 1.3 is *not* polynomial-time constructible.[4] We shall be much more interested in non-trivial cases of computational indistinguishability in which both ensembles are polynomial-time constructible. An important example is provided by the definition of pseudorandom generators (see Exercise 2.6). As we shall see (in Theorem 2.14), the existence of one-way functions implies the existence of pseudorandom generators, which in turn implies the existence of *polynomial-time constructible* probability ensembles that are statistically far apart and yet are computationally indistinguishable. We mention that this sufficient condition is also necessary (see Exercise 2.8).

2.3.3 Indistinguishability by multiple samples

The definition of computational indistinguishability (i.e., Definition 2.4) refers to distinguishers that obtain a single sample from one of the two relevant probability ensembles (i.e., $\{X_n\}_{n\in\mathbb{N}}$ and $\{Y_n\}_{n\in\mathbb{N}}$). A very natural generalization of Definition 2.4 refers to distinguishers that obtain several independent samples from such an ensemble.

[4] We say that $\{Z_n\}_{n\in\mathbb{N}}$ is polynomial-time constructible if there exists a polynomial-time algorithm S such that $S(1^n)$ and Z_n are identically distributed.

Definition 2.5 (indistinguishability by multiple samples): *Let $s : \mathbb{N} \rightarrow \mathbb{N}$ be polynomially-bounded. Two probability ensembles, $\{X_n\}_{n\in\mathbb{N}}$ and $\{Y_n\}_{n\in\mathbb{N}}$, are* computationally indistinguishable by $s(\cdot)$ samples *if for every probabilistic polynomial-time algorithm, D, every positive polynomial $p(\cdot)$, and all sufficiently large n, it holds that*

$$\left|\Pr\left[D(X_n^{(1)},...,X_n^{(s(n))})=1\right] - \Pr\left[D(Y_n^{(1)},...,Y_n^{(s(n))})=1\right]\right| < \frac{1}{p(n)}$$

where $X_n^{(1)}$ through $X_n^{(s(n))}$ and $Y_n^{(1)}$ through $Y_n^{(s(n))}$ are independent random variables such that each $X_n^{(i)}$ is identical to X_n and each $Y_n^{(i)}$ is identical to Y_n.

It turns out that, in the most interesting cases, computational indistinguishability by a single sample implies computational indistinguishability by any polynomial number of samples. One such case is the case of polynomial-time constructible ensembles. We say that the ensemble $\{Z_n\}_{n\in\mathbb{N}}$ is polynomial-time constructible if there exists a polynomial-time algorithm S such that $S(1^n)$ and Z_n are identically distributed (i.e., when given the parameter n (in unary), algorithm S produces a sample of Z_n is poly(n)-time).

Proposition 2.6 (indistinguishability is preserved under multiple samples): *Suppose that $X \stackrel{\text{def}}{=} \{X_n\}_{n\in\mathbb{N}}$ and $Y \stackrel{\text{def}}{=} \{Y_n\}_{n\in\mathbb{N}}$ are both polynomial-time constructible, and s is a positive polynomial. Then, X and Y are computationally indistinguishable by a single sample if and only if they are computationally indistinguishable by $s(\cdot)$ samples.*

Clearly, for every polynomial $s \geq 1$, computational indistinguishability by $s(\cdot)$ samples implies computational indistinguishability by a single sample (see Exercise 2.4). We now prove that, for efficiently constructible ensembles, indistinguishability by a single sample implies indistinguishability by multiple samples.[5] The proof provides a simple demonstration of a central proof technique, known as the *hybrid technique*, which is a special case of the so-called reducibility argument (cf. e.g., [22, Sec. 2.3.3] or [24, Sec. 7.1.2]).

Proof Sketch:[6] Using the counterpositive, we show that the existence of an efficient algorithm that distinguishes the ensembles X and Y using several samples, implies the existence of an efficient algorithm that distinguishes the ensembles X and Y using a single sample. That is, starting from the distinguishability of $s(n)$-long sequences of samples (either drawn all from X_n or drawn all from Y_n), we consider *hybrid* sequences such that the ith hybrid consists of i samples of X_n followed by $s(n) - i$ samples of Y_n. Note that the "homogeneous" sequences (which we assumed to be distinguishable) are the extreme hybrids (i.e., the first and last hybrids). The key observation is that distinguishing the extreme hybrids (towards the contradiction hypothesis) implies distinguishing neighboring hybrids, which in turn yields a procedure for distinguishing single samples of the two original distributions (contradicting the hypothesis that these two distributions are indistinguishable by a single sample). Details follow.

[5]The requirement that both ensembles are polynomial-time constructible is essential; see Exercise 2.9.

[6]For more details see [22, Sec. 3.2.3].

Suppose, towards the contradiction, that D distinguishes $s(n)$ samples of X_n from $s(n)$ samples of Y_n, with a distinguishing gap of $\delta(n)$. Denoting the ith hybrid by H_n^i (i.e., $H_n^i = (X_n^{(1)}, ..., X_n^{(i)}, Y_n^{(i+1)}, ..., Y_n^{(s(n))})$), this means that D distinguishes the extreme hybrids (i.e., H_n^0 and $H_n^{s(n)}$) with gap $\delta(n)$. It follows that D distinguishes a random pair of neighboring hybrids (i.e., D distinguishes H_n^i from H_n^{i+1}, for a randomly selected i) with gap at least $\delta(n)/s(n)$; the reason being that

$$\begin{aligned} &\mathsf{E}_{i\in\{0,...,s(n)-1\}}\left[\mathsf{Pr}[D(H_n^i)=1]-\mathsf{Pr}[D(H_n^{i+1})=1]\right] \\ &= \frac{1}{s(n)}\cdot\sum_{i=0}^{s(n)-1}\left(\mathsf{Pr}[D(H_n^i)=1]-\mathsf{Pr}[D(H_n^{i+1})=1]\right) \qquad (2.5) \\ &= \frac{1}{s(n)}\cdot\left(\mathsf{Pr}[D(H_n^0)=1]-\mathsf{Pr}[D(H_n^{s(n)})=1]\right) = \frac{\delta(n)}{s(n)}. \end{aligned}$$

The key step in the argument is transforming the distinguishability of neighboring hybrids into distinguishability of single samples of the original ensembles (thus deriving a contradiction). Indeed, using D, we obtain a distinguisher D' of single samples: Given a single sample, algorithm D' selects $i \in \{0, ..., s(n)-1\}$ at random, generates i samples from the first distribution and $s(n)-i-1$ samples from the second distribution, invokes D with the $s(n)$-samples sequence obtained when placing the input sample in location $i+1$, and answers whatever D does. That is, on input z and when selecting the index i, algorithm D' invokes D on a sample from the distribution $(X_n^{(1)}, ..., X_n^{(i)}, z, Y_n^{(i+2)}, ..., Y_n^{(s(n))})$. Thus, the construction of D' relies on the hypothesis that both probability ensembles are polynomial-time constructible. The analysis of D' is based on the following two facts:

1. When invoked on an input that is distributed according to X_n and selecting the index $i \in \{0, ..., s(n)-1\}$, algorithm D' behaves like $D(H_n^{i+1})$, because $(X_n^{(1)}, ..., X_n^{(i)}, X_n, Y_n^{(i+2)}, ..., Y_n^{(s(n))}) \equiv H_n^{i+1}$.

2. When invoked on an input that is distributed according to Y_n and selecting the index $i \in \{0, ..., s(n)-1\}$, algorithm D' behaves like $D(H_n^i)$, because $(X_n^{(1)}, ..., X_n^{(i)}, Y_n, Y_n^{(i+2)}, ..., Y_n^{(s(n))}) \equiv H_n^i$.

Thus, the distinguishing gap of D' (between Y_n and X_n) is captured by Eq. (2.5), and the claim follows. □

The hybrid technique – a digest: The hybrid technique constitutes a special type of a "reducibility argument" in which the computational indistinguishability of *complex* ensembles is proved using the computational indistinguishability of *basic* ensembles. The actual reduction is in the other direction: efficiently distinguishing the basic ensembles is reduced to efficiently distinguishing the complex ensembles, and *hybrid* distributions are used in the reduction in an essential way. The following three properties of the construction of the hybrids play an important role in the argument:

1. *The complex ensembles collide with the extreme hybrids.* This property is essential because our aim is to prove something that relates to the complex ensembles

(i.e., their indistinguishability), while the argument itself refers to the extreme hybrids.

In the proof of Proposition 2.6 the extreme hybrids (i.e., $H_n^{s(n)}$ and H_n^0) collide with the complex ensembles that represent $s(n)$-ary sequences of samples of one of the basic ensembles.

2. *The basic ensembles are efficiently mapped to neighboring hybrids.* This property is essential because our starting hypothesis relates to the basic ensembles (i.e., their indistinguishability), while the argument itself refers directly to the neighboring hybrids. Thus, we need to translate our knowledge (i.e., computational indistinguishability) of the basic ensembles to knowledge (i.e., computational indistinguishability) of any pair of neighboring hybrids. Typically, this is done by efficiently transforming strings in the range of a basic distribution into strings in the range of a hybrid such that the transformation maps the first basic distribution to one hybrid and the second basic distribution to the neighboring hybrid.

   In the proof of Proposition 2.6 the basic ensembles (i.e., X_n and Y_n) were efficiently transformed into neighboring hybrids (i.e., H_n^{i+1} and H_n^i, respectively). Recall that, in this case, the efficiency of this transformation relied on the hypothesis that both of the basic ensembles are polynomial-time constructible.

3. *The number of hybrids is small* (i.e., polynomial). This property is essential in order to deduce the computational indistinguishability of extreme hybrids from the computational indistinguishability of each pair of neighboring hybrids. Typically, the "distinguishability gap" established in the argument loses a factor that is proportional to the number of hybrids. This is due to the fact that the gap between the extreme hybrids is upper-bounded by the sum of the gaps between neighboring hybrids.

   In the proof of Proposition 2.6 the number of hybrids equals $s(n)$ and the aforementioned loss is reflected in Eq. (2.5).

We remark that in the course of an hybrid argument, a distinguishing algorithm referring to the complex ensembles is being analyzed and even invoked on arbitrary hybrids. The reader may be annoyed by the fact that the algorithm "was not designed to work on such hybrids" (but rather only on the extreme hybrids). However, *an algorithm is an algorithm*: once it exists we can invoke it on inputs of our choice, and analyze its performance on arbitrary input distributions.

2.4 Amplifying the Stretch Function

Recall that the definition of pseudorandom generators (i.e., Definition 2.1) makes a minimal requirement regarding their stretch; that is, it is only required that the output of such generators is longer than their input. Needless to say, we seek pseudorandom generators with a much more significant stretch, because the stretch determines the saving in randomness obtained in applications (e.g., via Construction 2.2). It turns out (see Construction 2.7) that pseudorandom generators of any stretch function (and in particular of minimal stretch $\ell_1(k) \stackrel{\text{def}}{=} k+1$) can be easily converted into pseudorandom generators of any desired (polynomially bounded) stretch function, ℓ.

On the other hand, since pseudorandom generators are required (by Definition 2.1) to run in polynomial time, their stretch must be polynomially bounded.

Construction 2.7 (stretch amplification): *Let G_1 be a pseudorandom generator with stretch function $\ell_1(k) = k+1$, and let ℓ be any polynomially bounded stretch function that is polynomial-time computable. Let*

$$G(s) \stackrel{\text{def}}{=} \sigma_1\sigma_2\cdots\sigma_{\ell(|s|)} \tag{2.6}$$

where $x_0 = s$ and $x_i\sigma_i = G_1(x_{i-1})$, for $i = 1, ..., \ell(|s|)$. That is, σ_i is the last bit of $G_1(x_{i-1})$ and x_i is the $|s|$-bit long prefix of $G_1(x_{i-1})$.

Needless to say, G is polynomial-time computable and has stretch ℓ. An alternative construction is considered in Exercise 2.10.

Proposition 2.8 (analysis of Construction 2.7): *Let G_1 and G be as in Construction 2.7. Then G constitutes a pseudorandom generator.*

Proof Sketch: The proposition is proven using the *hybrid technique*, presented and discussed in Section 2.3. Here (for $i = 0, ..., \ell(k)$) we consider the hybrid distributions H_k^i defined by

$$H_k^i \stackrel{\text{def}}{=} U_i^{(1)} \cdot g_{\ell(k)-i}(U_k^{(2)}),$$

where $\cdot$ denotes the concatenation of strings, $g_j(x)$ denotes the j-bit long prefix of $G(x)$, and $U_i^{(1)}$ and $U_k^{(2)}$ are independent uniform distributions (over $\{0,1\}^i$ and $\{0,1\}^k$, respectively). The extreme hybrids (i.e., H_k^0 and H_k^k) correspond to $G(U_k)$ and $U_{\ell(k)}$, whereas distinguishability of neighboring hybrids can be worked into distinguishability of $G_1(U_k)$ and U_{k+1}. Details follow.

Suppose that one could distinguish H_k^i from H_k^{i+1}. Defining $F(z)$ (resp., $L(z)$) as the first $|z|-1$ bits (resp., last bit) of z, and using $g_j(s) = L(G_1(s)) \cdot g_{j-1}(F(G_1(s)))$ (for $j \geq 1$), we have

$$H_k^i \equiv U_i^{(1)} \cdot L(G_1(U_k^{(2)})) \cdot g_{(\ell(k)-i)-1}(F(G_1(U_k^{(2)})))$$

and

$$\begin{aligned} H_k^{i+1} &= U_{i+1}^{(1')} \cdot g_{\ell(k)-(i+1)}(U_k^{(2)}) \\ &\equiv U_i^{(1)} \cdot L(U_{k+1}^{(2')}) \cdot g_{(\ell(k)-i)-1}(F(U_{k+1}^{(2')})). \end{aligned}$$

Now, incorporating the generation of $U_i^{(1)}$ and the evaluation of $g_{\ell(k)-i-1}$ into the distinguisher, it follows that we distinguish $G_1(U_k^{(2)})$ from $U_{k+1}^{(2')}$, in contradiction to the pseudorandomness of G_1. For further details see [24, Sec. 8.2.4] (or [22, Sec. 3.3.3]). □

Conclusion. In view of the foregoing, when talking about the mere existence of pseudorandom generators, in the sense of Definition 2.1, we may ignore the specific stretch function.

2.5 Constructions

So far we have ignored the basic question of whether pseudorandom generators exist at all. (Needless to say, the fact that we defined an object does not mean that it exists.) Looking at the definition of pseudorandom generators, we may observe that it implies the computational difficulty of inverting the transformation of seeds to longer output sequences (see details following Theorem 2.14). Thus, the existence of functions that are easy to compute but hard to invert, called *one-way functions*, is a necessary condition to the existence of pseudorandom generators, Interestingly, this condition is also sufficient; that is, pseudorandom generators can be constructed based on any one-way function.

We note that proving the equivalence of two seemingly different conditions is particularly beneficial when one of the two conditions seems simpler than the other and/or when we have more intuition regarding its validity. In particular, the conjectured existence of one-way functions is supported by the conjectured infeasibility of several well-known computational problems (e.g., factoring integers, decoding random linear codes, and computing logarithms in various finite groups).

2.5.1 Background: one-way functions

One-way functions are functions that are easy to compute but hard to invert (in an average-case sense).

Definition 2.9 (one-way functions): *A function $f:\{0,1\}^*\to\{0,1\}^*$ is called* one-way *if the following two conditions hold:*

1. Easy to evaluate: *There exists a polynomial-time algorithm A such that $A(x) = f(x)$ for every $x \in \{0,1\}^*$.*

2. Hard to invert: *For every probabilistic polynomial-time algorithm A', every positive polynomial p, and all sufficiently large n, it holds that*

$$\mathsf{Pr}_{x\in\{0,1\}^n}[A'(f(x),1^n) \in f^{-1}(f(x))] < \frac{1}{p(n)} \tag{2.7}$$

where the probability is taken uniformly over the possible choices of $x \in \{0,1\}^n$ and over the internal coin tosses of algorithm A'.

Algorithm A' is given the auxiliary input 1^n so as to allow it to run in time polynomial in the length of x, which is important in case f drastically shrinks its input (e.g., $|f(x)| = O(\log|x|)$). Typically (and, in fact, without loss of generality), the function f is length preserving, in which case the auxiliary input 1^n is redundant. Note that A' is not required to output a specific preimage of $f(x)$; any preimage (i.e., element in the set $f^{-1}(f(x))$) will do. (Indeed, in case f is one-to-one, the string x is the only preimage of $f(x)$ under f; but in general there may be other preimages.) It is required that algorithm A' fails (to find a preimage) with overwhelming probability, when the probability is also taken over the input distribution. That is, f is "typically" hard to invert, not merely hard to invert in some ("rare") cases.

On hard-core predicates. Recall that saying that a function f is one-way means that given a typical y (in the range of f) it is infeasible to find a preimage of y under f. This does not mean that it is infeasible to find partial information about the preimage(s) of y under f. Specifically, it may be easy to retrieve half of the bits of the preimage (e.g., given a one-way function f consider the function f' defined by $f'(x,r) \stackrel{\text{def}}{=} (f(x),r)$, for every $|x|=|r|$). We note that hiding partial information (about the function's preimage) plays an important role in the construction of pseudorandom generators (as well as in other advanced constructs). With this motivation in mind, we will show that essentially any one-way function hides specific partial information about its preimage, where this partial information is easy to compute from the preimage itself. This partial information can be considered a "hard-core" of the difficulty of inverting f. Loosely speaking, a *polynomial-time computable* (Boolean) predicate b, is called a hard-core of a function f if no feasible algorithm, given $f(x)$, can guess $b(x)$ with success probability that is non-negligibly better than one half.

Definition 2.10 (hard-core predicates): *A polynomial-time computable predicate $b : \{0,1\}^* \to \{0,1\}$ is called a* hard-core *of a function f if for every probabilistic polynomial-time algorithm A', every positive polynomial $p(\cdot)$, and all sufficiently large n, it holds that*

$$\mathsf{Pr}_{x\in\{0,1\}^n}\left[A'(f(x))=b(x)\right] < \frac{1}{2} + \frac{1}{p(n)}$$

where the probability is taken uniformly over the possible choices of $x \in \{0,1\}^n$ and over the internal coin tosses of algorithm A'.

Note that for every $b : \{0,1\}^* \to \{0,1\}$ and $f : \{0,1\}^* \to \{0,1\}^*$, there exist obvious algorithms that guess $b(x)$ from $f(x)$ with success probability at least one half (e.g., the algorithm that, obliviously of its input, outputs a uniformly chosen bit). Also, if b is a hard-core predicate (of any function), then it follows that b is almost unbiased (i.e., for a uniformly chosen x, the difference $|\mathsf{Pr}[b(x)=0] - \mathsf{Pr}[b(x)=1]|$ must be a negligible function in n).

Since b itself is polynomial-time computable, the failure of efficient algorithms to approximate $b(x)$ from $f(x)$ (with success probability that is non-negligibly higher than one half) must be due either to an information loss of f (i.e., f not being one-to-one) or to the difficulty of inverting f. For example, for $\sigma\in\{0,1\}$ and $x'\in\{0,1\}^*$, the predicate $b(\sigma x') = \sigma$ is a hard-core of the function $f(\sigma x') \stackrel{\text{def}}{=} 0x'$. Hence, in this case the fact that b is a hard-core of the function f is due to the fact that f loses information (specifically, the first bit: σ). On the other hand, in the case that f loses no information (i.e., f is one-to-one) a hard-core for f may exist only if f is hard to invert. In general, the interesting case is when being a hard-core is a computational phenomenon rather than an information theoretic one (which is due to "information loss" of f). It turns out that any one-way function has a modified version that possesses a hard-core predicate.

Theorem 2.11 (a generic hard-core predicate): *For any one-way function f, the inner-product mod 2 of x and r, denoted $b(x,r)$, is a hard-core of $f'(x,r) = (f(x),r)$.*

In other words, Theorem 2.11 asserts that, given $f(x)$ and a random subset $S \subseteq \{1,...,|x|\}$, it is infeasible to guess $\bigoplus_{i\in S} x_i$ significantly better than with probability

1/2, where $x = x_1 \cdots x_n$ is uniformly distributed in $\{0,1\}^n$. The proof of Theorem 2.11 appears in Appendix C.[7]

2.5.2 A simple construction

Intuitively, the definition of a hard-core predicate implies a potentially interesting case of computational indistinguishability. Specifically, as will be shown implicitly in Proposition 2.12 and explicitly in Exercise 2.7, if b is a hard-core of the function f, then the ensemble $\{f(U_n) \cdot b(U_n)\}_{n\in\mathbb{N}}$ is computationally indistinguishable from the ensemble $\{f(U_n) \cdot U'_1\}_{n\in\mathbb{N}}$. Furthermore, if f is one-to-one then the foregoing ensembles are statistically far apart, and thus constitute a non-trivial case of computational indistinguishability. If f is also polynomial-time computable and length-preserving, then this yields a construction of a pseudorandom generator.

Proposition 2.12 (A simple construction of pseudorandom generators): *Let b be a hard-core predicate of a polynomial-time computable one-to-one and length-preserving function f. Then, $G(s) \stackrel{\text{def}}{=} f(s) \cdot b(s)$ is a pseudorandom generator.*

Proof Sketch: Considering a uniformly distributed $s \in \{0,1\}^n$, we first note that the n-bit long prefix of $G(s)$ is uniformly distributed in $\{0,1\}^n$, because f induces a permutation on the set $\{0,1\}^n$. Hence, the proof boils down to showing that distinguishing $f(s) \cdot b(s)$ from $f(s) \cdot \sigma$, where σ is a random bit, yields a contradiction to the hypothesis that b is a hard-core of f (i.e., that $b(s)$ is *unpredictable* from $f(s)$). Intuitively, the reason is that such a hypothetical distinguisher also distinguishes $f(s) \cdot b(s)$ from $f(s) \cdot \overline{b(s)}$, where $\overline{\sigma} = 1 - \sigma$, whereas distinguishing $f(s) \cdot b(s)$ from $f(s) \cdot \overline{b(s)}$ yields an algorithm for predicting $b(s)$ based on $f(s)$. For further details see [24, Sec. 8.2.5.1] (or [22, Sec. 3.3.4]). □

Combining Theorem 2.11, Proposition 2.12 and Construction 2.7, we obtain the following result.

Theorem 2.13 (a sufficient condition for the existence of pseudorandom generators): *If there exist one-to-one and length-preserving one-way functions, then, for every polynomially bounded stretch function ℓ, there exists a pseudorandom generator of stretch ℓ.*

Digest. The main part of the proof of Proposition 2.12 is showing that the (next-bit) unpredictability of $G(U_k)$ implies the pseudorandomness of $G(U_k)$. The fact that (next-bit) unpredictability and pseudorandomness are equivalent, in general, is proven explicitly in the alternative proof of Theorem 2.13 provided next.

2.5.3 An alternative presentation

Let us take a closer look at the pseudorandom generators obtained by combining Construction 2.7 and Proposition 2.12. For a stretch function $\ell : \mathbb{N} \to \mathbb{N}$, a one-to-one

[7] We provide this proof because, in our opinion, at the last account, the conversion of computational difficulty to pseudorandomness occurs in this result. In contrast, important results such as Propositions 2.8 and 2.12 "only" transform one type of pseudorandomness into another. On the other hand, the proof of Theorem 2.11 is too long to fit in the main text without damaging the main thread of the presentation.

one-way function f with a hard-core b, we obtain

$$G(s) \stackrel{\text{def}}{=} \sigma_1\sigma_2\cdots\sigma_{\ell(|s|)}, \tag{2.8}$$

where $x_0 = s$ and $x_i\sigma_i = f(x_{i-1})b(x_{i-1})$ for $i = 1, ..., \ell(|s|)$. Denoting by $f^i(x)$ the value of f iterated i times on x (i.e., $f^i(x) = f^{i-1}(f(x))$ and $f^0(x) = x$), we rewrite Eq. (2.8) as follows:

$$G(s) \stackrel{\text{def}}{=} b(s)\cdot b(f(s))\cdots b(f^{\ell(|s|)-1}(s)). \tag{2.9}$$

The pseudorandomness of G is established in two steps, using the notion of (next-bit) unpredictability. An ensemble $\{Z_k\}_{k\in\mathbb{N}}$ is called unpredictable if any probabilistic polynomial-time machine obtaining a (random)[8] prefix of Z_k fails to predict the next bit of Z_k with probability non-negligibly higher than $1/2$. Specifically, we establish the following two results.

1. A **general result** asserting that *an ensemble is pseudorandom if and only if it is unpredictable.* Recall that an ensemble is pseudorandom if it is computationally indistinguishable from a uniform distribution (over bit strings of adequate length).

   Clearly, pseudorandomness implies polynomial-time unpredictability, but here we actually need the other direction, which is less obvious. Still, using a hybrid argument, one can show that (next-bit) unpredictability implies indistinguishability from the uniform ensemble. For details see Exercise 2.11.

2. A **specific result** asserting that the ensemble $\{G(U_k)\}_{k\in\mathbb{N}}$ is unpredictable *from right to left.* Equivalently, $G'(U_n)$ is polynomial-time unpredictable (from left to right (as usual)), where $G'(s) = b(f^{\ell(|s|)-1}(s))\cdots b(f(s))\cdot b(s)$ is the reverse of $G(s)$.

   Using the fact that f induces a permutation over $\{0,1\}^n$, observe that the $(j+1)$-bit long prefix of $G'(U_k)$ is distributed identically to $b(f^j(U_k))\cdots b(f(U_k))\cdot b(U_k)$. Thus, an algorithm that predicts the $j+1^{\text{st}}$ bit of $G'(U_n)$ based on the j-bit long prefix of $G'(U_n)$ yields an algorithm that guesses $b(U_n)$ based on $f(U_n)$. For details see Exercise 2.13.

Needless to say, G is a pseudorandom generator if and only if G' is a pseudorandom generator (see Exercise 2.12). We mention that Eq. (2.9) is often referred to as the Blum-Micali Construction.[9]

2.5.4 A necessary and sufficient condition

Recall that given any one-way one-to-one length-preserving function, we can easily construct a pseudorandom generator. Actually, the one-to-one (and length-preserving) requirement may be dropped, but the currently known construction – for the general case – is quite complex.

[8] For simplicity, we define unpredictability as referring to prefixes of a random length (distributed uniformly in $\{0, ..., |Z_k|-1\}$). A more general definition allows the predictor to determine the length of the prefix that it reads on the fly. This seemingly stronger notion of unpredictability is actually equivalent to the one we use, because both notions are equivalent to pseudorandomness.

[9] Given the popularity of the term, we deviate from our convention of not specifying credits in the main text. Indeed, this construction originates in [11].

Theorem 2.14 (on the existence of pseudorandom generators): *Pseudorandom generators exist if and only if one-way functions exist.*

To show that the existence of pseudorandom generators imply the existence of one-way functions, consider a pseudorandom generator G with stretch function $\ell(k) = 2k$. For $x, y \in \{0,1\}^k$, define $f(x,y) \stackrel{\text{def}}{=} G(x)$, and so f is polynomial-time computable (and length-preserving). It must be that f is one-way, or else one can distinguish $G(U_k)$ from U_{2k} by trying to invert f and checking the result: inverting f on the distribution $f(U_{2k})$ corresponds to operating on the distribution $G(U_k)$, whereas the probability that U_{2k} has an inverse under f is negligible.

The interesting direction of the proof of Theorem 2.14 is the construction of pseudorandom generators based on any one-way function. Since the known proof is quite complex, we only provide a very rough overview of some of the ideas involved. We mention that these ideas make extensive use of adequate hashing functions (e.g., pairwise independent hashing functions; see Appendix A).

We first note that, in general (when f may not be one-to-one), the ensemble $f(U_k)$ may not be pseudorandom, and so the construction of Proposition 2.12 (i.e., $G(s) = f(s)b(s)$, where b is a hard-core of f) cannot be used *directly*. One idea underlying the known construction is hashing $f(U_k)$ to an almost uniform string of length that almost equals its entropy.[10] But "hashing $f(U_k)$ down to length comparable to the entropy" means shrinking the length of the output to, say, $k' < k$. This foils the entire point of stretching the k-bit seed. Thus, a second idea underlying the construction is compensating for the loss of $k - k'$ bits by extracting these many bits from the seed U_k itself. This is done by hashing U_k, and the point is that the $(k-k')$-bit long hash value does not make the inverting task any easier. Implementing these ideas turns out to be more difficult than it seems, and indeed an alternative construction would be most appreciated.

2.6 Non-uniformly Strong Pseudorandom Generators

Recall that we said that truly random sequences can be replaced by pseudorandom sequences without affecting any efficient computation that uses these sequences. The specific formulation of this assertion, presented in Proposition 2.3, refers to randomized algorithms that take a "primary input" and use a secondary "random input" in their computation. Proposition 2.3 asserts that it is infeasible to find a primary input for which the replacement of a truly random secondary input by a pseudorandom one affects the final output of the algorithm in a noticeable way. This, however, does not mean that such primary inputs do not exist (but rather that they are hard to find). Consequently, Proposition 2.3 falls short of yielding a (worst-case)[11] "deran-

[10]This is done after guaranteeing that the logarithm of the probability mass of a value of $f(U_k)$ is typically close to the entropy of $f(U_k)$. Specifically, given an arbitrary one-way function f', one first constructs f by taking a "direct product" of sufficiently many copies of f'. For example, for $x_1, ..., x_{k^{2/3}} \in \{0,1\}^{k^{1/3}}$, we let $f(x_1, ..., x_{k^{2/3}}) \stackrel{\text{def}}{=} f'(x_1), ..., f'(x_{k^{2/3}})$.

[11]Indeed, Proposition 2.3 yields an *average-case derandomization of $\mathcal{BPP}$*. In particular, for every polynomial-time constructible ensemble $\{X_n\}_{n\in\mathbb{N}}$, every Boolean function $f \in \mathcal{BPP}$, and every $\varepsilon > 0$, there exists a randomized algorithm A' of randomness complexity $r_\varepsilon(n) = n^\varepsilon$ such that the probability that $A'(X_n) \neq f(X_n)$ is negligible. A corresponding deterministic ($\exp(r_\varepsilon)$-

domization" of a complexity class such as $\mathcal{BPP}$. To obtain such results, we need a stronger notion of pseudorandom generators, presented next. Specifically, we need pseudorandom generators that can fool all polynomial-size circuits, and not merely all probabilistic polynomial-time algorithms.[12]

Definition 2.15 (strong pseudorandom generator – fooling circuits): *A deterministic polynomial-time algorithm G is called a* non-uniformly strong pseudorandom generator *if there exists a* stretch function, $\ell:\mathbb{N}\to\mathbb{N}$, *such that for any family $\{C_k\}_{k\in\mathbb{N}}$ of polynomial-size circuits, for any positive polynomial p, and for all sufficiently large k, it holds that*

$$|\,\mathsf{Pr}[C_k(G(U_k)) = 1] \;-\; \mathsf{Pr}[C_k(U_{\ell(k)}) = 1]\,| \;<\; \frac{1}{p(k)}\,.$$

Using such pseudorandom generators, we can "derandomize" $\mathcal{BPP}$.

Theorem 2.16 (derandomization of $\mathcal{BPP}$): *If there exist non-uniformly strong pseudorandom generators, then $\mathcal{BPP}$ is contained in $\bigcap_{\varepsilon>0}\mathrm{DTIME}(t_\varepsilon)$, where $t_\varepsilon(n) \stackrel{\rm def}{=} 2^{n^\varepsilon}$.*

See Appendix F for definitions of the aforementioned complexity classes.

Proof Sketch: For any $S \in \mathcal{BPP}$ and any $\varepsilon > 0$, we let A denote a probabilistic polynomial-time decision procedure for S and let G denote a non-uniformly strong pseudorandom generator stretching n^ε-bit long seeds into poly(n)-long sequences (to be used by A as secondary input when processing a primary input of length n). Combining A and G, we obtain an algorithm $A' = A_G$ (as in Construction 2.2). We claim that *A and A' may significantly differ in their* (expected probabilistic) *decision on at most finitely many inputs*, because otherwise we can use these inputs (together with A) to derive a (non-uniform) family of polynomial-size circuits that distinguishes $G(U_{n^\varepsilon})$ and $U_{\mathrm{poly}(n)}$, contradicting the the hypothesis regarding G. Specifically, an input x on which A and A' differ significantly yields a circuit C_x that distinguishes $G(U_{|x|^\varepsilon})$ and $U_{\mathrm{poly}(|x|)}$, by letting $C_x(r) = A(x,r)$.[13] Incorporating the finitely many "bad" inputs into A', we derive a probabilistic polynomial-time algorithm that decides S while using randomness complexity n^ε.

Finally, emulating A' on each of the 2^{n^ε} possible random sequences (i.e., seeds to G) and ruling by majority, we obtain a deterministic algorithm A'' as required. That is, let $A'(x,r)$ denote the output of algorithm A' on input x when using coins $r \in \{0,1\}^{n^\varepsilon}$. Then $A''(x)$ invokes $A'(x,r)$ on every $r \in \{0,1\}^{n^\varepsilon}$, and outputs 1 if and only if the majority of these 2^{n^ε} invocations have returned 1. □

time) algorithm A'' can be obtained, as in the proof of Theorem 2.16, and again the probability that $A''(X_n) \neq f(X_n)$ is negligible, where here the probability is taken only over the distribution of the primary input (represented by X_n). In contrast, worst-case derandomization, as captured by the assertion $\mathcal{BPP} \subseteq \mathrm{DTIME}(2^{r_\varepsilon})$, requires that the probability that $A''(X_n) \neq f(X_n)$ is zero.

[12]Needless to say, strong pseudorandom generators in the sense of Definition 2.15 satisfy the basic definition of a pseudorandom generator (i.e., Definition 2.1); see Exercise 2.14. We comment that the underlying notion of computational indistinguishability (by circuits) is strictly stronger than Definition 2.4, and that it is invariant under multiple samples (regardless of the constructibility of the underlying ensembles); for details, see Exercise 2.15.

[13]Indeed, in terms of the proof of Proposition 2.3, the finder F consists of a non-uniform family of polynomial-size circuits that print the "problematic" primary inputs that are hard-wired in them, and the corresponding distinguisher D is thus also non-uniform.

We comment that stronger results regarding derandomization of $\mathcal{BPP}$ are presented in Section 3.

On constructing non-uniformly strong pseudorandom generators. Non-uniformly strong pseudorandom generators (as in Definition 2.15) can be constructed using any one-way function that is hard to invert by any non-uniform family of polynomial-size circuits, rather than by probabilistic polynomial-time machines. In fact, the construction in this case is simpler than the one employed in the uniform case (i.e., the construction underlying the proof of Theorem 2.14).

2.7 Stronger (Uniform-Complexity) Notions

The following two notions represent strengthening of the standard definition of pseudorandom generators (as presented in Definition 2.1). Non-uniform versions of these notions (strengthening Definition 2.15) are also of interest.

2.7.1 Fooling stronger distinguishers

One strengthening of Definition 2.1 amounts to explicitly quantifying the resources (and success gaps) of distinguishers. We choose to bound these quantities as a function of the length of the seed (i.e., k), rather than as a function of the length of the string that is being examined (i.e., $\ell(k)$). For a class of time bounds $\mathcal{T}$ (e.g., $\mathcal{T} = \{t(k) \stackrel{\text{def}}{=} 2^{c\sqrt{k}}\}_{c\in\mathbb{N}}$) and a class of noticeable functions (e.g., $\mathcal{F} = \{f(k) \stackrel{\text{def}}{=} 1/t(k) : t \in \mathcal{T}\}$), we say that a pseudorandom generator, G, is $(\mathcal{T},\mathcal{F})$-strong if for any probabilistic algorithm D having running-time bounded by a function in $\mathcal{T}$ (applied to k)[14], for any function f in $\mathcal{F}$, and for all sufficiently large k's, it holds that

$$|\Pr[D(G(U_k)) = 1] - \Pr[D(U_{\ell(k)}) = 1]| < f(k).$$

An analogous strengthening may be applied to the definition of one-way functions. Doing so reveals the weakness of the known construction that underlies the proof of Theorem 2.14; it only implies that for some $\varepsilon > 0$ ($\varepsilon = 1/8$ will do), for any $\mathcal{T}$ and $\mathcal{F}$, the existence of "$(\mathcal{T},\mathcal{F})$-strong one-way functions" implies the existence of $(\mathcal{T}',\mathcal{F}')$-strong pseudorandom generators, where $\mathcal{T}' = \{t'(k) \stackrel{\text{def}}{=} t(k^{\varepsilon})/\text{poly}(k) : t \in \mathcal{T}\}$ and $\mathcal{F}' = \{f'(k) \stackrel{\text{def}}{=} \text{poly}(k) \cdot f(k^{\varepsilon}) : f \in \mathcal{F}\}$. What we *would like* to have is an analogous result with $\mathcal{T}' = \{t'(k) \stackrel{\text{def}}{=} t(\Omega(k))/\text{poly}(k) : t \in \mathcal{T}\}$ and $\mathcal{F}' = \{f'(k) \stackrel{\text{def}}{=} \text{poly}(k) \cdot f(\Omega(k)) : f \in \mathcal{F}\}$.

2.7.2 Pseudorandom functions

Recall that pseudorandom *generators* provide a way to efficiently generate long pseudorandom sequences from short random seeds. Pseudorandom functions are even more powerful: they provide efficient direct access to the bits of a huge pseudorandom sequence (which is not feasible to scan bit-by-bit). More precisely, a pseudorandom function is an efficient (deterministic) algorithm that given a k-bit *seed*, s, and a

[14]That is, when examining a sequence of length $\ell(k)$ algorithm D makes at most $t(k)$ steps, where $t \in \mathcal{T}$.

k-bit *argument*, x, returns a k-bit string, denoted $f_s(x)$, such that it is infeasible to distinguish the values of f_s, for a uniformly chosen $s \in \{0,1\}^k$, from the values of a truly random function $F : \{0,1\}^k \to \{0,1\}^k$. That is, the (feasible) testing procedure is given oracle access to the function (but not its explicit description), and cannot distinguish the case when it is given oracle access to a pseudorandom function from the case when it is given oracle access to a truly random function.

Definition 2.17 (pseudorandom functions): *A* pseudorandom function (ensemble), *is a collection of functions* $\{f_s : \{0,1\}^{|s|} \to \{0,1\}^{|s|}\}_{s \in \{0,1\}^*}$ *that satisfies the following two conditions:*

1. (efficient evaluation) *There exists an efficient* (deterministic) *algorithm that given a* seed, *s, and an* argument, $x \in \{0,1\}^{|s|}$, *returns* $f_s(x)$.

2. (pseudorandomness) *For every probabilistic polynomial-time oracle machine, M, every positive polynomial p and all sufficiently large k, it holds that*

$$\left| \Pr[M^{f_{U_k}}(1^k) = 1] - \Pr[M^{F_k}(1^k) = 1] \right| < \frac{1}{p(k)}$$

where F_k *denotes a uniformly selected function mapping* $\{0,1\}^k$ *to* $\{0,1\}^k$.

One key feature of pseudorandom functions is that they can be generated and shared by merely generating and sharing their seed; that is, a "random looking" function $f_s : \{0,1\}^k \to \{0,1\}^k$, is determined by its k-bit seed s. Thus, parties wishing to share a "random looking" function f_s (determining 2^k-many values), merely need to generate and share among themselves the k-bit seed s. (For example, one party may randomly select the seed s, and communicate it, via a secure channel, to all other parties.) Sharing a pseudorandom function allows parties to determine (by themselves and without any further communication) random-looking values depending on their current views of the environment (which need not be known a priori). To appreciate the potential of this tool, one should realize that sharing a pseudorandom function is essentially as good as being able to agree, on the fly, on the association of random values to (on-line) given values, where the latter are taken from a huge set of possible values. We stress that this agreement is achieved without communication and synchronization: Whenever some party needs to associate a random value to a given value, $v \in \{0,1\}^k$, it will associate to v the (same) random value $r_v \in \{0,1\}^k$ (by setting $r_v = f_s(v)$, where f_s is a pseudorandom function agreed upon beforehand). Concretely, the foregoing idea underlies the construction of secure private-key encryption and message-authentication schemes based on pseudorandom functions (see Appendix E). In addition to numerous applications in cryptography, pseudorandom functions were also used to derive negative results in computational learning theory [69] and in the study of circuit complexity (cf., Natural Proofs [55]).

Theorem 2.18 (how to construct pseudorandom functions): *Pseudorandom functions can be constructed using any pseudorandom generator.*

Proof Sketch:[15] Let G be a pseudorandom generator that stretches its seed by a factor of two (i.e., $\ell(k) = 2k$), and let $G_0(s)$ (resp., $G_1(s)$) denote the first (resp.,

[15] See details in [22, Sec. 3.6.2].

last) $|s|$ bits in $G(s)$. Let

$$G_{\sigma_{|s|}\cdots\sigma_2\sigma_1}(s) \stackrel{\text{def}}{=} G_{\sigma_{|s|}}(\cdots G_{\sigma_2}(G_{\sigma_1}(s))\cdots),$$

define $f_s(x_1x_2\cdots x_k) \stackrel{\text{def}}{=} G_{x_k\cdots x_2x_1}(s)$, and consider the function ensemble $\{f_s : \{0,1\}^{|s|} \to \{0,1\}^{|s|}\}_{s\in\{0,1\}^*}$. Pictorially, the function f_s is defined by k-step walks down a full binary tree of depth k having labels at the vertices. The root of the tree, hereafter referred to as the level 0 vertex of the tree, is labeled by the string s. If an internal vertex is labeled r, then its left child is labeled $G_0(r)$ whereas its right child is labeled $G_1(r)$. The value of $f_s(x)$ is the string residing in the leaf reachable from the root by a path corresponding to the string x.

We claim that the function ensemble $\{f_s\}_{s\in\{0,1\}^*}$ is pseudorandom. The proof uses the hybrid technique (cf. Section 2.3): The ith hybrid, denoted H_k^i, is a function ensemble consisting of $2^{2^i\cdot k}$ functions $\{0,1\}^k \to \{0,1\}^k$, each determined by 2^i random k-bit strings, denoted $\overline{s} = \langle s_\beta\rangle_{\beta\in\{0,1\}^i}$. The value of such a function $h_{\overline{s}}$ at $x = \alpha\beta$, where $|\beta| = i$, is defined to equal $G_\alpha(s_\beta)$. Pictorially, the function $h_{\overline{s}}$ is defined by placing the strings in $\overline{s}$ in the corresponding vertices of level i, and labeling vertices of lower levels using the very rule used in the definition of f_s. The extreme hybrids correspond to our indistinguishability claim (i.e., $H_k^0 \equiv f_{U_k}$ and H_k^k is a truly random function), and the indistinguishability of neighboring hybrids follows from our indistinguishability hypothesis. Specifically, we show that the ability to distinguish H_k^i from H_k^{i+1} yields an ability to distinguish multiple samples of $G(U_k)$ from multiple samples of U_{2k} (by placing on the fly, halves of the given samples at adequate vertices of the $i+1^{\text{st}}$ level). □

Variants. Useful variants (and generalizations) of the notion of pseudorandom functions include Boolean pseudorandom functions that are defined over all strings (i.e., $f_s : \{0,1\}^* \to \{0,1\}$) and pseudorandom functions that are defined for other domains and ranges (i.e., $f_s : \{0,1\}^{d(|s|)} \to \{0,1\}^{r(|s|)}$, for arbitrary polynomially bounded functions $d, r : \mathbb{N} \to \mathbb{N}$). Various transformations between these variants are known (cf. [22, Sec. 3.6.4] and [23, Apdx. C.2]).

2.8 Conceptual Reflections

We highlight several conceptual aspects of the foregoing computational approach to randomness. Some of these aspects are common to other instantiation of the general paradigm (esp., the one presented in Chapter 3).

Behavioristic versus Ontological. The behavioristic nature of the computational approach to randomness is best demonstrated by confronting this approach with the Kolmogorov-Chaitin approach to randomness. Loosely speaking, a string is *Kolmogorov-random* if its length equals the length of the shortest program producing it. This shortest program may be considered the "true explanation" to the phenomenon described by the string. A Kolmogorov-random string is thus a string that does not have a substantially simpler (i.e., shorter) explanation than itself. Considering the simplest explanation of a phenomenon may be viewed as an ontological approach. In contrast, considering the effect of phenomena on certain devices (or observations), as underlying the definition of pseudorandomness, is a behavioristic approach. Furthermore, there exist probability distributions that are not uniform (and are not even statistically close to a uniform distribution) and, nevertheless, are indistinguishable from a uniform distribution (by any efficient device). Thus, *distributions that are ontologically very different, are considered equivalent by the behavioristic point of view taken in the definition of computational indistinguishability.*

A relativistic view of randomness. We have defined pseudorandomness in terms of its observer. Specifically, we have considered the class of efficient (i.e., polynomial-time) observers and defined as pseudorandom objects that look random to any observer in that class. In subsequent chapters, we shall consider restricted classes of such observers (e.g., space-bounded polynomial-time observers and even very restricted observers that merely apply specific tests such as linear tests or hitting tests). Each such class of observers gives rise to a different notion of pseudorandomness. Furthermore, the general paradigm (of pseudorandomness) explicitly aims at *distributions that are not uniform and yet are considered as such from the point of view of certain observers.* Thus, our entire approach to pseudorandomness is relativistic and subjective (i.e., depending on the abilities of the observer).

Randomness and Computational Difficulty. Pseudorandomness and computational difficulty play dual roles: The general paradigm of pseudorandomness relies on the fact that *placing computational restrictions on the observer gives rise to distributions that are not uniform and still cannot be distinguished from uniform distributions.* Thus, the pivot of the entire approach is the computational difficulty of distinguishing pseudorandom distributions from truly random ones. Furthermore, many of the constructions of pseudorandom generators rely either on conjectures or on facts regarding computational difficulty (i.e., that certain computations are hard for certain classes). For example, one-way functions were used to construct general-purpose pseudorandom generators (i.e., those working in polynomial-time and fooling all polynomial-time observers). Analogously, as we shall see in Section 3.2.3, the fact that parity function is hard for polynomial-size constant-depth circuits can be used to generate (highly non-uniform) sequences that fool such circuits.

Randomness and Predictability. The connection between pseudorandomness and unpredictability (by efficient procedures) plays an important role in the analysis of several constructions (cf. Sections 2.5 and 3.2). Here, we wish to highlight the intuitive appeal of this connection.

Notes

The concept of *computational indistinguishability*, which underlies the entire computational approach to randomness, was suggested by Goldwasser and Micali [29] in the context of defining secure encryption schemes. Indeed, computational indistinguishability plays a key role in cryptography (see [22, 23]). The general formulation of computational indistinguishability is due to Yao [73]. Using the hybrid technique of [29], Yao also observed that defining pseudorandom generators as producing sequences that are computationally indistinguishable from the corresponding uniform distribution is equivalent to defining such generators as producing unpredictable sequences. The latter definition originates in the earlier work of Blum and Micali [11].

Blum and Micali [11] pioneered the rigorous study of pseudorandom generators and, in particular, the construction of pseudorandom generators based on some simple intractability assumption. In particular, they constructed pseudorandom generators assuming the intractability of the Discrete Logarithm Problem (over prime fields). Their work also introduces basic paradigms that were used in all subsequent improvements (cf., e.g., [73, 32]). We refer to the transformation of computational difficulty into pseudorandomness, the use of hard-core predicates (also defined in [11]), and the iteration paradigm (cf. Eq. (2.9)).

Theorem 2.14 (by which pseudorandom generators exist if and only if one-way functions exist) is due to Håstad, Impagliazzo, Levin and Luby [32], building on the hard-core predicate of [27] (see Theorem 2.11). Unfortunately, the current proof of Theorem 2.14 is very complicated and unfit for presentation in this primer. Presenting a simpler and tighter (cf. Section 2.7) proof is indeed an important research project.

Pseudorandom functions were defined and first constructed by Goldreich, Goldwasser and Micali [25]. We also mention (and advocate) the study of a general theory of pseudorandom objects initiated in [26]. Finally, we mention that a more detailed treatment of general-purpose pseudorandom generators is provided in [22, Chap. 3].

Exercises

Exercise 2.1 Prove the following corollaries to Proposition 2.3.

1. Let A be a probabilistic polynomial-time algorithm solving a decision problem $\chi : \{0,1\}^* \to \{0,1\}$ (in $\mathcal{BPP}$), and let A_G be as in Construction 2.2. Prove that it is infeasible to find an x on which A_G errs with probability that is significantly higher than the error probability of A; that is, prove that on input 1^n it is infeasible to find $x \in \{0,1\}^n$ such that $\Pr[A_G(x) \neq \chi(x)] < \Pr[A(x) = \chi(x)] + 0.01$.

2. Let A be a probabilistic polynomial-time algorithm solving the search associated with the NP-relation R, and let A_G be as in Construction 2.2. Prove that it is infeasible to find an x on which A_G outputs a wrong solution; that is, assuming for simplicity that A has error probability $1/3$, prove that on input 1^n it is infeasible to find $x \in \{0,1\}^n \cap S_R$ such that $\Pr[(x, A_G(x)) \notin R] > 0.4$, where $S_R \stackrel{\text{def}}{=} \{x : \exists y\, (x,y) \in R\}$. Likewise, it is infeasible to find $x \in \{0,1\}^n \setminus S_R$ such that $\Pr[A_G(x) \neq \bot] > 0.4$.

Exercise 2.2 Prove that omitting the absolute value in Eq. (2.4) keeps Definition 2.4 intact.
(Hint: Consider $D'(z) \stackrel{\text{def}}{=} 1 - D(z)$.)

Exercise 2.3 Prove that computational indistinguishability is an equivalence relation (defined over pairs of probability ensembles). Specifically, prove that this relation is transitive (i.e., $X \equiv Y$ and $Y \equiv Z$ implies $X \equiv Z$).

Exercise 2.4 Prove that *if $\{X_n\}_{n\in\mathbb{N}}$ and $\{Y_n\}_{n\in\mathbb{N}}$ are computationally indistinguishable and A is a probabilistic polynomial-time algorithm, then $\{A(X_n)\}_{n\in\mathbb{N}}$ and $\{A(Y_n)\}_{n\in\mathbb{N}}$ are computationally indistinguishable.*

Guideline: If D distinguishes the latter ensembles, then D' such that $D'(z) \stackrel{\text{def}}{=} D(A(z))$ distinguishes the former.

Exercise 2.5 In contrast to Exercise 2.4, show that the conclusion may not hold when A is not computationally bounded. That is, show that there exists computationally indistinguishable ensembles, $\{X_n\}_{n\in\mathbb{N}}$ and $\{Y_n\}_{n\in\mathbb{N}}$, and an exponential-time algorithm A such that $\{A(X_n)\}_{n\in\mathbb{N}}$ and $\{A(Y_n)\}_{n\in\mathbb{N}}$ are *not* computationally indistinguishable.

Guideline: For any pair of ensembles $\{X_n\}_{n\in\mathbb{N}}$ and $\{Y_n\}_{n\in\mathbb{N}}$, consider the Boolean function f such that $f(z) = 1$ if and only if $\Pr[X_n = z] > \Pr[Y_n = z]$. Show that $|\Pr[f(X_n) = 1] - \Pr[f(Y_n) = 1]|$ equals the statistical difference between X_n and Y_n. Consider an adequate (approximate) implementation of f (e.g., approximate $\Pr[X_n = z]$ and $\Pr[Y_n = z]$ up to $\pm 2^{-2|z|}$).

Exercise 2.6 Show that the existence of pseudorandom generators implies the existence of polynomial-time constructible probability ensembles that are statistically far apart and yet are computationally indistinguishable.

Guideline: Lower-bound the statistical distance between $G(U_k)$ and $U_{\ell(k)}$, where G is a pseudorandom generator with stretch ℓ.

Exercise 2.7 Relying on Theorem 2.11, provide a self-contained proof of the fact that the existence of one-way one-to-one functions implies the existence of polynomial-time constructible probability ensembles that are statistically far apart and yet are computationally indistinguishable.

Guideline: Assuming that b is a hard-core of the function f, consider the ensembles $\{f(U_n) \cdot b(U_n)\}_{n\in\mathbb{N}}$ and $\{f(U_n) \cdot U'_1\}_{n\in\mathbb{N}}$. Prove that these ensembles are computationally indistinguishable by using the main ideas of the proof of Proposition 2.12. Show that if f is one-to-one, then these ensembles are statistically far apart.

Exercise 2.8 (following [20]) Prove that the sufficient condition in Exercise 2.6 is in fact necessary. Recall that $\{X_n\}_{n\in\mathbb{N}}$ and $\{Y_n\}_{n\in\mathbb{N}}$ are said to be statistically far apart if, for some positive polynomial p and all sufficiently large n, the variation distance between X_n and Y_n is greater than $1/p(n)$. Using the following three steps, prove that the existence of *polynomial-time constructible* probability ensembles that are statistically far apart and yet are computationally indistinguishable implies the existence of pseudorandom generators.

1. Show that, without loss of generality, we may assume that the variation distance between X_n and Y_n is greater than $1 - \exp(-n)$.

   Guideline: For X_n and Y_n as in the foregoing, consider $\overline{X}_n = (X_n^{(1)}, ..., X_n^{(t(n))})$ and $\overline{Y}_n = (Y_n^{(1)}, ..., Y_n^{(t(n))})$, where the $X_n^{(i)}$'s (resp., $Y_n^{(i)}$'s) are independent copies of X_n (resp., Y_n), and $t(n) = O(n \cdot p(n)^2)$. To lower-bound the statistical difference between $\overline{X}_n$ and $\overline{Y}_n$, consider the set $S_n \stackrel{\text{def}}{=} \{z : \Pr[X_n = z] > \Pr[Y_n = z]\}$ and the random variable representing the number of copies in $\overline{X}_n$ (resp., $\overline{Y}_n$) that reside in S_n.

2. Using $\{X_n\}_{n\in\mathbb{N}}$ and $\{Y_n\}_{n\in\mathbb{N}}$ as in Step 1, prove the existence of a *false entropy generator*, where a false entropy generator is a deterministic polynomial-time algorithm G such that $G(U_k)$ has entropy $e(k)$ but $\{G(U_k)\}_{k\in\mathbb{N}}$ is computationally indistinguishable from a polynomial-time constructible ensemble that has entropy greater than $e(\cdot) + (1/2)$.

   Guideline: Let S_0 and S_1 be sampling algorithms such that $X_n \equiv S_0(U_{\text{poly}(n)})$ and $Y_n \equiv S_1(U_{\text{poly}(n)})$. Consider the generator $G(\sigma, r) = (\sigma, S_\sigma(r))$, and the distribution Z_n that equals (U_1, X_n) with probability 1/2 and (U_1, Y_n) otherwise. Note that in $G(U_1, U_{\text{poly}(n)})$ the first bit is almost determined by the rest, whereas in Z_n the first bit is statistically independent of the rest.

3. Using a false entropy generator, obtain one in which the excess entropy is $\sqrt{k}$, and using the latter construct a pseudorandom generator.

   Guideline: Use the ideas presented in Section 2.5.4 (i.e., the discussion of the interesting direction of the proof of Theorem 2.14).

Exercise 2.9 (multiple samples vs. single sample, a separation) In contrast to Proposition 2.6, prove that there exist two probability ensembles that are computational indistinguishable by a single sample, but are efficiently distinguishable by two samples. Furthermore, one of these ensembles is the uniform ensemble and the other has a sparse support (i.e., only poly(n) many strings are assigned a non-zero probability weight by the second distribution). Indeed, the second ensemble is not polynomial-time constructible.

Guideline: Prove that, for every function $d : \{0,1\}^n \to [0,1]$, there exists two strings, x_n and y_n (in $\{0,1\}^n$), and a number $p \in [0,1]$ such that $\Pr[d(U_n)=1] = p \cdot \Pr[d(x_n)=1] + (1-p) \cdot \Pr[d(y_n)=1]$. Generalize this claim to m functions, using $m+1$ strings and a convex combination of the corresponding probabilities.[16] Conclude that there exists a distribution Z_n with a support of size at most $m+1$ such that for each of the first (in lexicographic order) m (randomized) algorithms A it holds that $\Pr[A(U_n) = 1] = \Pr[A(Z_n) = 1]$. Note that with probability at least $1/(m+1)$, two independent samples of Z_n are assigned the same value, yielding a simple two-sample distinguisher of U_n from Z_n.

Exercise 2.10 (amplifying the stretch function, an alternative) For G_1 and ℓ as in Construction 2.7, consider $G(s) \stackrel{\text{def}}{=} G_1^{\ell(|s|)-|s|}(s)$, where $G_1^i(x)$ denotes G_1 iterated i times on x (i.e., $G_1^i(x) = G_1^{i-1}(G_1(x))$ and $G_1^0(x) = x$). Prove that G is a pseudorandom generator of stretch ℓ. Reflect on the advantages of Construction 2.7 over the current construction (e.g., consider generation time).

Guideline: Use a hybrid argument, with the ith hybrid being $G_1^i(U_{\ell(k)-i})$, for $i = 0, ..., \ell(k) - k$. Note that $G_1^{i+1}(U_{\ell(k)-(i+1)}) = G_1^i(G_1(U_{\ell(k)-i-1}))$ and $G_1^i(U_{\ell(k)-i}) = G_1^i(U_{|G_1(U_{\ell(k)-i-1})|})$, and use Exercise 2.4.

Exercise 2.11 (pseudorandom vs. unpredictability) Prove that a probability ensemble $\{Z_k\}_{k\in\mathbb{N}}$ is pseudorandom if and only if it is unpredictable. For simplicity,

---

[16] That is, prove that for every m functions $d_1, ..., d_m : \{0,1\}^n \to [0,1]$ there exist $m+1$ strings $z_n^{(1)}, ..., z_n^{(m+1)}$ and $m+1$ non-negative numbers $p_1, ..., p_{m+1}$ that sum-up to 1 such that for every $i \in \{1, ..., m\}$ it holds that $\Pr[d_i(U_n)=1] = \sum_j p_j \cdot \Pr[d_i(z_n^{(j)})=1]$.

we say that $\{Z_k\}_{k\in\mathbb{N}}$ is (next-bit) unpredictable if for every probabilistic polynomial-time algorithm A it holds that $\mathsf{Pr}_i[A(F_i(Z_k))=B_{i+1}(Z_k)]-(1/2)$ is negligible, where $i \in \{0,...,|Z_k|-1\}$ is uniformly distributed, and $F_i(z)$ (resp., $B_{i+1}(z)$) denotes the i-bit prefix (resp., $i+1^{\rm st}$ bit) of z.

Guideline: Show that pseudorandomness implies polynomial-time unpredictability; that is, polynomial-time predictability violates pseudorandomness (because the uniform ensemble is unpredictable regardless of computing power). Use a hybrid argument to prove that unpredictability implies pseudorandomness. Specifically, the $i^{\rm th}$ hybrid consists of the i-bit long prefix of Z_k followed by $|Z_k|-i$ uniformly distributed bits. Thus, distinguishing the extreme hybrids (which correspond to Z_k and $U_{|Z_k|}$) implies distinguishing a random pair of neighboring hybrids, which in turn implies next-bit predictability. For the last step, use an argument as in the proof of Proposition 2.12.

Exercise 2.12 Prove that a probability ensemble is unpredictable (from left to right) if and only if it is unpredictable from right to left (or in any other canonical order).

Guideline: Use Exercise 2.11, and note that an ensemble is pseudorandom if and only if its reverse is pseudorandom.

Exercise 2.13 Let f be one-to-one and length preserving, and let b be a hard-core predicate of f. For any polynomial ℓ, letting $G'(s) \stackrel{\rm def}{=} b(f^{\ell(|s|)-1}(s))\cdots b(f(s))\cdot b(s)$, prove that $\{G'(U_k)\}$ is unpredictable (in the sense of Exercise 2.11).

Guideline: Suppose towards the contradiction that, for a uniformly distributed $j \in \{0,...,\ell(k)-1\}$, given the j-bit long prefix of $G'(U_k)$ an algorithm A' can predict the $j+1^{\rm st}$ bit of $G'(U_k)$. That is, given $b(f^{\ell(k)-1}(s))\cdots b(f^{\ell(k)-j}(s))$, algorithm A' predicts $b(f^{\ell(k)-(j+1)}(s))$, where s is uniformly distributed in $\{0,1\}^k$. Consider an algorithm A that given $y=f(x)$ approximates $b(x)$ by invoking A' on input $b(f^{j-1}(y))\cdots b(y)$, where j is uniformly selected in $\{0,...,\ell(k)-1\}$. Analyze the success probability of A using the fact that f induces a permutation over $\{0,1\}^n$, and thus $b(f^j(U_k))\cdots b(f(U_k))\cdot b(U_k)$ is distributed identically to $b(f^{\ell(k)-1}(U_k))\cdots b(f^{\ell(k)-j}(U_k))\cdot b(f^{\ell(k)-(j+1)}(U_k))$.

Exercise 2.14 Prove that if G is a strong pseudorandom generator in the sense of Definition 2.15, then it a pseudorandom generator in the sense of Definition 2.1.

Guideline: Consider a sequence of internal coin tosses that maximizes the probability in Eq. (2.1).

Exercise 2.15 (strong computational indistinguishability) Provide a definition of the notion of computational indistinguishability that underlies Definition 2.15 (i.e., indistinguishability with respect to (non-uniform) polynomial-size circuits). Prove the following two claims:

1. Computational indistinguishability with respect to (non-uniform) polynomial-size circuits is strictly stronger than Definition 2.4.

2. Computational indistinguishability with respect to (non-uniform) polynomial-size circuits is invariant under (polynomially-many) multiple samples, even if the underlying ensembles are not polynomial-time constructible.

Guideline: For Part 1, see the solution to Exercise 2.9. For Part 2 note that samples as generated in the proof of Proposition 2.6 can be hard-wired into the distinguishing circuit.

Chapter 3

Derandomization of Time-Complexity Classes

Let us take a second look at the process of derandomization that underlies the proof of Theorem 2.16. First, a pseudorandom generator was used to shrink the randomness-complexity of a BPP-algorithm, and then derandomization was achieved by scanning all possible seeds to this generator. A key observation regarding this process is that there is no point in insisting that the pseudorandom generator runs in time that is polynomial in its seed length. Instead, it suffices to require that the generator runs in time that is exponential in its seed length, because we are already incurring such an overhead due to the scanning of all possible seeds. Furthermore, in this context, the running-time of the generator may be larger than the running time of the algorithm, which means that the generator need only fool distinguishers that take fewer steps than the generator. These considerations motivate the following definition of canonical derandomizers.

3.1 Defining Canonical Derandomizers

Recall that in order to "derandomize" a probabilistic polynomial-time algorithm A, we first obtain a functionally equivalent algorithm A_G (as in Construction 2.2) that has (significantly) smaller randomness-complexity. Algorithm A_G has to maintain A's input-output behavior on all (but finitely many) inputs. Thus, the set of the relevant distinguishers (considered in the proof of Theorem 2.16) is the set of all possible circuits obtained from A by hard-wiring any of the possible inputs. Such a circuit, denoted C_x, emulates the execution of algorithm A on input x, when using the circuit's input as the algorithm's internal coin tosses (i.e., $C_x(r) = A(x,r)$). Furthermore, the size of C_x is quadratic in the running-time of A on input x, and the length of the input to C_x equals the running-time of A (on input x).[1] Thus,

[1] Indeed, we assume that algorithm A is represented as a Turing machine and refer to the standard emulation of Turing machines by circuits. Thus, the aforementioned circuit C_x has size that is at most quadratic in the running-time of A on input x, which in turn means that C_x has size that is at most quadratic in the length of its own input. (In fact, the circuit size can be made almost-linear in the running-time of A, by using a better emulation [54].) We note that many sources use the fictitious convention by which the circuit size equals the length of its input; this fictitious convention

the size of C_x is quadratic in the length of its own input, and the pseudorandom generator in use (i.e., G) needs to fool each such circuit. Recalling that we may allow the generator to run in exponential-time (i.e., time that is exponential in the length of its own input (i.e., the seed))[2], we arrive at the following definition.

Definition 3.1 (pseudorandom generator for derandomizing BPTIME($\cdot$))[3]: *Let $\ell : \mathbb{N} \to \mathbb{N}$ be a monotonically increasing function. A* canonical derandomizer of stretch ℓ *is a deterministic algorithm G that satisfies the following two conditions.*

1. *On input a k-bit long seed, G makes at most* $\text{poly}(2^k \cdot \ell(k))$ *steps and outputs a string of length $\ell(k)$.*

2. *For every circuit D_k of size $\ell(k)^2$ it holds that*

$$|\,\mathsf{Pr}[D_k(G(U_k)) = 1] - \mathsf{Pr}[D_k(U_{\ell(k)}) = 1]\,| < \frac{1}{6}. \tag{3.1}$$

The circuit D_k represents a potential distinguisher, which is given an $\ell(k)$-bit long string (sampled either from $G(U_k)$ or from $U_{\ell(k)}$). When seeking to derandomize an algorithm A of time-complexity t, the aforementioned $\ell(k)$-bit long string represents a possible sequence of coin tosses of A, when invoked on a generic (primary) input of length $n = t^{-1}(\ell(k))$. Thus, for any $x \in \{0,1\}^n$, considering the circuit $D_k(r) = A(x,r)$, where $|r| = t(n) = \ell(k)$, we note that Eq. (3.1) implies that $A_G(x) = A(x, G(U_k))$ *maintains the majority vote of* $A(x) = A(x, U_{\ell(k)})$. On the other hand, the time-complexity of G implies that the straightforward deterministic emulation of $A_G(x)$ takes time $2^k \cdot (\text{poly}(2^k \cdot \ell(k)) + t(n))$, which is upper-bounded by $\text{poly}(2^k \cdot \ell(k)) = \text{poly}(2^{\ell^{-1}(t(n))} \cdot t(n))$. This yields the following (conditional) derandomization result.

Proposition 3.2 (using canonical derandomizers): *Let $\ell, t : \mathbb{N} \to \mathbb{N}$ be monotonically increasing functions and let $\ell^{-1}(t(n))$ denote the smallest integer k such that $\ell(k) \geq t(n)$. If there exists a canonical derandomizer of stretch ℓ, then, for every time-constructible $t : \mathbb{N} \to \mathbb{N}$, it holds that* $\text{BPTIME}(t) \subseteq \text{DTIME}(T)$, *where* $T(n) = \text{poly}(2^{\ell^{-1}(t(n))} \cdot t(n))$.

Proof Sketch: Just mimic the proof of Theorem 2.16, which in turn uses Construction 2.2. (Recall that given any randomized algorithm A and generator G, Construction 2.2 yields an algorithm A_G of randomness-complexity $\ell^{-1} \circ t$ and time-complexity

can be justified by considering a (suitably) padded input.

[2]Actually, in Definition 3.1 we allow the generator to run in time $\text{poly}(2^k \ell(k))$, rather than in time $\text{poly}(2^k)$. This is done in order not to trivially rule out generators of super-exponential stretch (i.e., $\ell(k) = 2^{\omega(k)}$). However (see Exercise 3.2), the condition in Eq. (3.1) does not allow for super-exponential stretch (or even for $\ell(k) = \omega(2^k)$). Thus, in retrospect, the two formulations are equivalent (because $\text{poly}(2^k \ell(k)) = \text{poly}(2^k)$ for $\ell(k) = 2^{O(k)}$).

[3]Fixing a model of computation, we denote by BPTIME(t) the class of decision problems that are solvable by a randomized algorithm of time complexity t that has a two-sided error probability of at most 1/3. Using 1/6 as the "threshold distinguishing gap" (in Eq. (3.1)) guarantees that if $\mathsf{Pr}[D_k(U_{\ell(k)}) = 1] \geq 2/3$ (resp., $\mathsf{Pr}[D_k(U_{\ell(k)}) = 1] \leq 1/3$), then $\mathsf{Pr}[D_k(G(U_k)) = 1] > 1/2$ (resp., $\mathsf{Pr}[D_k(G(U_k)) = 1] < 1/2$). As we shall see, this suffices for a derandomization of BPTIME(t) in time T, where $T(n) = \text{poly}(2^{\ell^{-1}(t(n))} \cdot t(n))$ (and we use a seed of length $k = \ell^{-1}(t(n))$).

$\text{poly}(2^{\ell^{-1} \circ t}) + t$.)[4] Observe that the complexity of the resulting deterministic procedure is dominated by the $2^k = 2^{\ell^{-1}(t(|x|))}$ invocations of $A_G(x, s) = A(x, G(s))$, where $s \in \{0,1\}^k$, and each of these invocations takes time $\text{poly}(2^{\ell^{-1}(t(|x|))}) + t(|x|)$. Thus, on input an n-bit long string, the deterministic procedure runs in time $\text{poly}(2^{\ell^{-1}(t(n))} \cdot t(n))$. The correctness of this procedure (which takes a majority vote among the 2^k invocations of A_G) follows by combining Eq. (3.1) with the hypothesis that $\mathsf{Pr}[A(x)=1]$ is bounded away from $1/2$. Specifically, using the hypothesis $|\mathsf{Pr}[A(x)=1] - (1/2)| \geq 1/6$, it follows that the majority vote of $(A_G(x,s))_{s \in \{0,1\}^k}$ equals 1 if and only if $\mathsf{Pr}[A(x)=1] > 1/2$. Indeed, the implication is due to Eq. (3.1), when applied to the circuit $C_x(r) = A(x, r)$ (which has size at most $|r|^2$). □

The goal. In light of Proposition 3.2, we seek canonical derandomizers with a stretch that is as large as possible. The stretch cannot be super-exponential (i.e., it must hold that $\ell(k) = O(2^k)$), because there exists a circuit of size $O(2^k \cdot \ell(k))$ that violates Eq. (3.1) (see Exercise 3.2) whereas for $\ell(k) = \omega(2^k)$ it holds that $O(2^k \cdot \ell(k)) < \ell(k)^2$. Thus, our goal is to construct a canonical derandomizer with stretch $\ell(k) = 2^{\Omega(k)}$. Such a canonical derandomizer will allow for a "full derandomization of $\mathcal{BPP}$":

Theorem 3.3 (derandomization of $\mathcal{BPP}$, revisited): *If there exists a canonical derandomizer of stretch $\ell(k) = 2^{\Omega(k)}$, then $\mathcal{BPP} = \mathcal{P}$.*

Proof: Using Proposition 3.2, we get $\text{BPTIME}(t) \subseteq \text{DTIME}(T)$, where $T(n) = \text{poly}(2^{\ell^{-1}(t(n))} \cdot t(n)) = \text{poly}(t(n))$. ■

Reflections: Recall that a canonical derandomizer G was defined in a way that allows it to have time-complexity t_G that is larger than the size of the circuits that it fools (i.e., $t_G(k) > \ell(k)^2$ is allowed). Furthermore, $t_G(k) > 2^k$ was also allowed. Thus, if indeed $t_G(k) = 2^{\Omega(k)}$ (as is the case in Section 3.2), then $G(U_k)$ *can be distinguished from* $U_{\ell(k)}$ *in time* $2^k \cdot t_G(k) = \text{poly}(t_G(k))$ by trying all possible seeds.[5] We stress that the latter distinguisher is a uniform algorithm (and it works by invoking G on all possible seeds). In contrast, for a general-purpose pseudorandom generator G (as discussed in Chapter 2) it holds that $t_G(k) = \text{poly}(k)$, while *for every polynomial p it holds that $G(U_k)$ is indistinguishable from $U_{\ell(k)}$ in time $p(t_G(k))$.*

3.2 Constructing Canonical Derandomizers

The fact that canonical derandomizers are allowed to be more complex than the corresponding distinguisher makes *some* of the techniques of Chapter 2 inapplicable

[4] Actually, given any randomized algorithm A and generator G, Construction 2.2 yields an algorithm A_G that is defined such that $A_G(x,s) = A(x, G'(s))$, where $|s| = \ell^{-1}(t(|x|))$ and $G'(s)$ denotes the $t(|x|)$-bit long prefix of $G(s)$. For simplicity, we shall assume here that $\ell(|s|) = t(|x|)$, and thus use G rather than G'. Note that given n we can find $k = \ell^{-1}(t(n))$ by invoking $G(1^i)$ for $i = 1, ..., k$ (using the fact that $\ell : \mathbb{N} \rightarrow \mathbb{N}$ is monotonically increasing). Also note that $\ell(k) = O(2^k)$ must hold (see Footnote 2), and thus we may replace $\text{poly}(2^k \cdot \ell(k))$ by $\text{poly}(2^k)$.

[5] We note that this distinguisher does not contradict the hypothesis that G is a canonical derandomizer, because $t_G(k) > \ell(k)$ definitely holds whereas $\ell(k) \leq 2^k$ typically holds (and so $2^k \cdot t_G(k) > \ell(k)^2$).

in the current context. For example, the stretch function cannot be amplified as in Section 2.4 (see Exercise 3.1). On the other hand, the techniques developed in the current section are inapplicable to Chapter 2. For example, the pseudorandomness of some canonical derandomizers (i.e., the generators of Construction 3.4) holds even when the potential distinguisher is given the seed itself. This amazing phenomenon capitalizes on the fact that the distinguisher's time-complexity does not allow for running the generator on the given seed.

3.2.1 The construction and its consequences

As in Section 2.5, the construction presented next transforms computational difficulty into pseudorandomness, except that here both computational difficulty and pseudorandomness are of a somewhat different form than in Section 2.5. Specifically, here we use Boolean predicates that are computable in exponential-time but are strongly inapproximable; that is, we assume *the existence of a Boolean predicate and constants* $c, \varepsilon > 0$ *such that for all but finitely many* m, *the* (residual) *predicate* $f : \{0,1\}^m \to \{0,1\}$ *is computable in time* 2^{cm} *but for any circuit* C *of size* $2^{\varepsilon m}$ *it holds that* $\Pr[C(U_m) = f(U_m)] < \frac{1}{2} + 2^{-\varepsilon m}$. (Needless to say, $\varepsilon < c$.) Such predicates exist under the assumption that the class $\mathcal{E}$ (where $\mathcal{E} = \bigcup_{c>0} \text{DTIME}(2^{c \cdot n})$) contains predicates of (almost-everywhere) exponential circuit complexity [34]. With these preliminaries, we turn to the construction of canonical derandomizers with exponential stretch.

Construction 3.4 (The Nisan-Wigderson Construction):[6] *Let* $f : \{0,1\}^m \to \{0,1\}$ *and* $S_1, ..., S_\ell$ *be a sequence of* m*-subsets of* $\{1, ..., k\}$. *Then, for* $s \in \{0,1\}^k$, *we let*

$$G(s) \stackrel{\text{def}}{=} f(s_{S_1}) \cdots f(s_{S_\ell}) \tag{3.2}$$

where s_S *denotes the projection of* s *on the bit locations in* $S \subseteq \{1, ..., |s|\}$; *that is, for* $s = \sigma_1 \cdots \sigma_k$ *and* $S = \{i_1, ..., i_m\}$ *such that* $i_1 < \cdots < i_m$, *we have* $s_S = \sigma_{i_1} \cdots \sigma_{i_m}$.

Letting k vary and $\ell, m : \mathbb{N} \to \mathbb{N}$ be functions of k, we wish G to be a canonical derandomizer and $\ell(k) = 2^{\Omega(k)}$. One (obvious) necessary condition for this to happen is that the sets must be distinct, and hence $m(k) = \Omega(k)$; consequently, f must be computable in exponential-time. Furthermore, the sequence of sets $S_1, ..., S_{\ell(k)}$ must be constructible in $\text{poly}(2^k)$-time. Intuitively, the function f should be strongly inapproximable, and furthermore it seems desirable to use a set system with relatively small pairwise intersections (because this restricts the overlap among the various inputs to which f is applied). Interestingly, these conditions are essentially sufficient.

Theorem 3.5 (analysis of Construction 3.4): *Let* $\alpha, \beta, \gamma, \varepsilon > 0$ *be constants satisfying* $\varepsilon > (2\alpha/\beta) + \gamma$, *and consider the functions* $\ell, m, T : \mathbb{N} \to \mathbb{N}$ *such that* $\ell(k) = 2^{\alpha k}$, $m(k) = \beta k$, *and* $T(n) = 2^{\varepsilon n}$. *Suppose that the following two conditions hold:*

1. *There exists an exponential-time computable function* $f : \{0,1\}^* \to \{0,1\}$ *such that for every family of* T*-size circuits* $\{C_n\}_{n \in \mathbb{N}}$ *and all sufficiently large* n *it holds that*

$$\Pr[C_n(U_n) \neq f(U_n)] \geq \frac{1}{2} + \frac{1}{T(n)}\,. \tag{3.3}$$

[6] Given the popularity of the term, we deviate from our convention of not specifying credits in the main text. This construction originates in [49, 52].

*In this case we say that f is T-*inapproximable.

2. *There exists an exponential-time computable function $S:\mathbb{N}\times\mathbb{N}\rightarrow 2^{\mathbb{N}}$ such that:*
   (a) *For every k and $i \in \{1,...,\ell(k)\}$, it holds that $S(k,i) \subseteq \{1,...,k\}$ and $|S(k,i)| = m(k)$.*
   (b) *For every k and $i \neq j$, it holds that $|S(k,i) \cap S(k,j)| \leq \gamma \cdot m(k)$.*

Then, using G as defined in Construction 3.4 with $S_i = S(k,i)$, yields a canonical derandomizer with stretch ℓ.

Before proving Theorem 3.5 we mention that, for any $\gamma > 0$, a function S as in Condition 2 does exist for some $m(k) = \Omega(k)$ and $\ell(k) = 2^{\Omega(k)}$; see Exercise 3.3. We also recall that T-inapproximable predicates do exist under the assumption that $\mathcal{E}$ has (almost-everywhere) exponential circuit complexity (see [34] or [24, Sec. 8.2.1]). Thus, combining such functions f and S and invoking Theorem 3.5, we obtain a canonical derandomizer with exponential stretch based on the assumption that $\mathcal{E}$ has (almost-everywhere) exponential circuit complexity. Combining this with Theorem 3.3, we get the first part of the following theorem.

Theorem 3.6 (derandomization of BPP, revisited):

1. *Suppose that $\mathcal{E}$ contains a decision problem that has almost-everywhere exponential circuit complexity* (i.e., there exists a constant $\varepsilon_0 > 0$ such that, for all but finitely many m's, any circuit that correctly decides this problem on $\{0,1\}^m$ has size at least $2^{\varepsilon_0 m}$). *Then, $\mathcal{BPP} = \mathcal{P}$.*

2. *Suppose that, for every polynomial p, the class $\mathcal{E}$ contains a decision problem that has circuit complexity that is almost-everywhere greater than p. Then $\mathcal{BPP}$ is contained in $\bigcap_{\varepsilon>0} \mathrm{DTIME}(t_\varepsilon)$, where $t_\varepsilon(n) \stackrel{\text{def}}{=} 2^{n^\varepsilon}$.*

Indeed, our focus is on Part 1, and Part 2 is stated for the sake of a wider perspective. Both parts are special cases of a more general statement that can be proved by using a generalization of Theorem 3.5 that refers to arbitrary functions $\ell, m, T : \mathbb{N} \rightarrow \mathbb{N}$ (instead of the exponential functions in Theorem 3.5) that satisfy $\ell(k)^2 + \widetilde{O}(\ell(k) \cdot 2^{m'(k)}) < T(m(k))$, where $m'(k)$ replaces $\gamma \cdot m(k)$. (For details, see Exercise 3.6.) We note that Part 2 of Theorem 3.6 supersedes Theorem 2.16. We also mention that, as in the case of general-purpose pseudorandom generators, the hardness hypothesis used in each part of Theorem 3.6 is necessary for the existence of a corresponding canonical derandomizer (see Exercise 3.8).

Additional comment. The two parts of Theorem 3.6 exhibit two extreme cases: Part 1 (often referred to as the "high end") assumes an extremely strong circuit lower-bound and yields "full derandomization" (i.e., $\mathcal{BPP} = \mathcal{P}$), whereas Part 2 (often referred to as the "low end") assumes an extremely weak circuit lower-bound and yields weak but meaningful derandomization. Intermediate results (relying on intermediate lower-bound assumptions) can be obtained analogous to Exercise 3.7, but tight trade-offs are obtained differently (cf., [67]).

3.2.2 Analyzing the construction (i.e., proof of Theorem 3.5)

Using the time-complexity upper-bounds on f and S, it follows that G can be computed in exponential time. Thus, our focus is on showing that $\{G(U_k)\}$ cannot be distinguished from $\{U_{\ell(k)}\}$ by circuits of size $\ell(k)^2$; specifically, that G satisfies Eq. (3.1). In fact, we will prove that this holds for $G'(s) = s \cdot G(s)$; that is, G fools such circuits even if they are given the seed as auxiliary input. (Indeed, these circuits are smaller than the running time of G, and so they cannot just evaluate G on the given seed.)

We start by presenting the intuition underlying the proof. As a warm-up suppose that the sets (i.e., $S(k,i)$'s) used in the construction are disjoint. In such a case (which is indeed impossible because $k < \ell(k) \cdot m(k)$), the pseudorandomness of $G(U_k)$ would follow easily from the inapproximability of f, because in this case G consists of applying f to non-overlapping parts of the seed (see Exercise 3.5). In the actual construction being analyzed here, the sets (i.e., $S(k,i)$'s) are not disjoint but have relatively small pairwise intersection, which means that G applies f on parts of the seed that have relatively small overlap. Intuitively, such small overlaps guarantee that the values of f on the corresponding inputs are "computationally independent" (i.e., having the value of f at some inputs $x_1, ..., x_i$ does not help in approximating the value of f at another input x_{i+1}). This intuition will be backed by showing that, when fixing all bits that do not appear in the target input (i.e., in x_{i+1}), the former values (i.e., $f(x_1), ..., f(x_i)$) can be computed at a relatively small computational cost. Thus, the values $f(x_1), ..., f(x_i)$ do not (significantly) facilitate the task of approximating $f(x_{i+1})$. With the foregoing intuition in mind, we now turn to the actual proof.

The actual proof employs a reducibility argument; that is, assuming towards the contradiction that G' does not fool some circuit of size $\ell(k)^2$, we derive a contradiction to the hypothesis that the predicate f is T-inapproximable. The argument utilizes the relation between pseudorandomness and unpredictability (cf. Section 2.5). Specifically, as detailed in Exercise 3.4, *any circuit that distinguishes $G'(U_k)$ from $U_{\ell(k)+k}$ with gap $1/6$, yields a next-bit predictor of similar size that succeeds in predicting the next bit with probability at least* $\frac{1}{2} + \frac{1}{6\ell'(k)} > \frac{1}{2} + \frac{1}{7\ell(k)}$, where the factor of $\ell'(k) = \ell(k) + k < (1 + o(1)) \cdot \ell(k)$ is introduced by the hybrid technique (cf. Eq. (2.5)). Furthermore, given the non-uniform setting of the current proof, we may fix a bit location $i+1$ for prediction, rather than analyzing the prediction at a random bit location. Indeed, $i \geq k$ must hold, because the first k bits of $G'(U_k)$ are uniformly distributed. In the rest of the proof, we transform the foregoing predictor into a circuit that approximates f better than allowed by the hypothesis (regarding the inapproximability of f).

Assuming that a small circuit C' can predict the $i + 1^{\text{st}}$ bit of $G'(U_k)$, when given the previous i bits, we construct a small circuit C for approximating $f(U_{m(k)})$ on input $U_{m(k)}$. The point is that the $i + 1^{\text{st}}$ bit of $G'(s)$ equals $f(s_{S(k,j+1)})$, where $j = i - k \geq 0$, and so C' approximates $f(s_{S(k,j+1)})$ based on $s, f(s_{S(k,1)}), ..., f(s_{S(k,j)})$, where $s \in \{0,1\}^k$ is uniformly distributed. Note that this is the type of thing that we are after, except that the circuit we seek may only get $s_{S(k,j+1)}$ as input.

The first observation is that C' maintains its advantage when we fix the best choice for the bits of s that are not at bit locations $S_{j+1} = S(k, j+1)$ (i.e., the bits

$s_{[k]\setminus S_{j+1}}$, where $[k] \stackrel{\text{def}}{=} \{1, ...k\}$). That is, by an averaging argument, it holds that

$$\max_{s' \in \{0,1\}^{k-m(k)}} \{\Pr_{s \in \{0,1\}^k}[C'(s, f(s_{S_1}), ..., f(s_{S_j})) = f(s_{S_{j+1}}) \mid s_{[k]\setminus S_{j+1}} = s']\}$$
$$\geq p' \stackrel{\text{def}}{=} \Pr_{s \in \{0,1\}^k}[C'(s, f(s_{S_1}), ..., f(s_{S_j})) = f(s_{S_{j+1}})].$$

Recall that by the hypothesis $p' > \frac{1}{2} + \frac{1}{7\ell(k)}$. Hard-wiring the fixed string s' into C', and letting $\pi(x)$ denote the (unique) string s satisfying $s_{S_{j+1}} = x$ and $s_{[k]\setminus S_{j+1}} = s'$, we obtain a circuit C'' that satisfies

$$\Pr_{x \in \{0,1\}^{m(k)}}[C''(x, f(\pi(x)_{S_1}), ..., f(\pi(x)_{S_j})) = f(x)] \geq p'.$$

The circuit C'' is almost what we seek. The only problem is that C'' gets as input not only x, but also $f(\pi(x)_{S_1}), ..., f(\pi(x)_{S_j})$, whereas we seek an approximator of $f(x)$ that only gets x.

The key observation is that each of the "missing" values $f(\pi(x)_{S_1}), ..., f(\pi(x)_{S_j})$ depend only on a relatively small number of the bits of x. This fact is due to the hypothesis that $|S_t \cap S_{j+1}| \leq \gamma \cdot m(k)$ for $t = 1, ..., j$, which means that $\pi(x)_{S_t}$ is an $m(k)$-bit long string in which $m_t \stackrel{\text{def}}{=} |S_t \cap S_{j+1}|$ bits are projected from x and the rest are projected from the *fixed* string s'. Thus, given x, the value $f(\pi(x)_{S_t})$ can be computed by a (trivial) circuit of size $\widetilde{O}(2^{m_t})$; that is, by a circuit implementing a look-up table on m_t bits. Using all these circuits (together with C''), we will obtain the desired approximator of f. Details follow.

We obtain the desired circuit, denoted C, that T-approximates f as follows. The circuit C depends on the index j and the string s' that are fixed as in the foregoing analysis. Recall that C incorporates ($\widetilde{O}(2^{\gamma \cdot |x|})$-size) circuits for computing $x \mapsto f(\pi(x)_{S_t})$, for $t = 1, ..., j$. On input $x \in \{0,1\}^{m(k)}$, the circuit C computes the values $f(\pi(x)_{S_1}), ..., f(\pi(x)_{S_j})$, invokes C'' on input x and these values, and outputs the answer as a guess for $f(x)$. That is,

$$C(x) = C''(x, f(\pi(x)_{S_1}), ..., f(\pi(x)_{S_j})) = C'(\pi(x), f(\pi(x)_{S_1}), ..., f(\pi(x)_{S_j})).$$

By the foregoing analysis, $\Pr_x[C(x) = f(x)] \geq p' > \frac{1}{2} + \frac{1}{7\ell(k)}$, which is lower-bounded by $\frac{1}{2} + \frac{1}{T(m(k))}$, because $T(m(k)) = 2^{\varepsilon m(k)} = 2^{\varepsilon \beta k} \gg 2^{2\alpha k} \gg 7\ell(k)$, where the first inequality is due to $\varepsilon > 2\alpha/\beta$ and the second inequality is due to $\ell(k) = 2^{\alpha k}$. The size of C is upper-bounded by $\ell(k)^2 + \ell(k) \cdot \widetilde{O}(2^{\gamma \cdot m(k)}) \ll \widetilde{O}(\ell(k)^2 \cdot 2^{\gamma \cdot m(k)}) = \widetilde{O}(2^{2\alpha \cdot (m(k)/\beta) + \gamma \cdot m(k)}) \ll T(m(k))$, where the last inequality is due to $T(m(k)) = 2^{\varepsilon m(k)} \gg \widetilde{O}(2^{(2\alpha/\beta) \cdot m(k) + \gamma \cdot m(k)})$ (which in turn uses $\varepsilon > (2\alpha/\beta) + \gamma$). Thus, we derived a contradiction to the hypothesis that f is T-inapproximable. This completes the proof of Theorem 3.5.

3.2.3 Construction 3.4 as a general framework

The Nisan–Wigderson Construction (i.e., Construction 3.4) is actually a general framework, which can be instantiated in various ways. Some of these instantiations, which are based on an abstraction of the construction as well as of its analysis, are briefly reviewed next.

We first note that the generator described in Construction 3.4 consists of a generic algorithmic scheme that can be instantiated with any predicate f. Furthermore, this algorithmic scheme, denoted G, is actually an *oracle machine* that makes (non-adaptive) queries to the function f, and thus the combination (of G and f) may be written as G^f. Likewise, the proof of pseudorandomness of G^f (i.e., the bulk of the proof of Theorem 3.5) is actually a general scheme that, for every f, yields a (non-uniform) oracle-aided circuit C that approximates f by using an oracle call to any distinguisher for G^f (i.e., C uses the distinguisher as a black-box). The circuit C does depend on f (but in a restricted way). Specifically, C contains look-up tables for computing functions obtained from f by fixing some of the input bits (i.e., look-up tables for the functions $f(\pi(\cdot)_{S_t})$'s). The foregoing abstractions facilitate the presentation of the following instantiations of the general framework underlying Construction 3.4

Derandomization of constant-depth circuits. In this case we instantiate Construction 3.4 using the `parity` function in the role of the inapproximable predicate f, noting that `parity` is indeed inapproximable by "small" constant-depth circuits.[7] With an adequate setting of parameters we obtain pseudorandom generators with stretch $\ell(k) = \exp(k^{1/O(1)})$ that fool "small" constant-depth circuits (see [49]). The analysis of this construction proceeds very much like the proof of Theorem 3.5. One important observation is that incorporating the (straightforward) circuits that compute $f(\pi(x)_{S_t})$ into the distinguishing circuit only increases its depth by two levels. Specifically, the circuit C uses depth-two circuits that compute the values $f(\pi(x)_{S_t})$'s, and then obtains a prediction of $f(x)$ by using these values in its (single) invocation of the (given) distinguisher.

The resulting pseudorandom generator, which uses a seed of polylogarithmic length (equiv., $\ell(k) = \exp(k^{1/O(1)})$), can be used for derandomizing $\mathcal{RAC}^0$ (i.e., random $\mathcal{AC}^0$)[8], analogously to Theorem 3.3. Thus, we can *deterministically* approximate, in quasi-polynomial-time and up to an additive error, the fraction of inputs that satisfy a given (constant-depth) circuit. Specifically, for any constant d, given a depth-d circuit C, we can deterministically approximate the fraction of the inputs that satisfy C (i.e., cause C to evaluate to 1) to within any *additive constant error*[9] in time $\exp((\log|C|)^{O(d)})$. Providing a deterministic polynomial-time approximation, even when $d = 2$ (i.e., CNF/DNF formulae) is an open problem.

Derandomization of probabilistic proof systems. A different (and more surprising) instantiation of Construction 3.4 utilizes predicates that are inapproximable by small *circuits having oracle access to $\mathcal{NP}$*. The result is a pseudorandom generator robust against two-move public-coin interactive proofs (which are as powerful as constant-round interactive proofs). The key observation is that the analysis of Construction 3.4 provides a black-box procedure for approximating the underlying predicate when given oracle access to a distinguisher (and this procedure is valid

[7] See references in [49].

[8] The class $\mathcal{AC}^0$ consists of all decision problems that are solvable by constant-depth circuits of polynomial size (and unbounded fan-in).

[9] We mention that in the special case of approximating the number of satisfying assignment of a DNF formula, *relative error* approximations can be obtained by employing a deterministic reduction of relative error approximation to additive constant error approximation (see [21, Apdx. B.1.1] or [24, §6.2.2.1]). Thus, using a pseudorandom generator that fools DNF formulae, we can deterministically obtain a relative (rather than additive) error approximation to the number of satisfying assignment in a given DNF formula.

also in case the distinguisher is a non-deterministic machine). Thus, under suitably strong (and yet plausible) assumptions, constant-round interactive proofs collapse to $\mathcal{NP}$. We note that a stronger result, which deviates from the foregoing framework, has been subsequently obtained (cf. [45]).

Construction of randomness extractors. An even more radical instantiation of Construction 3.4 was used to obtain explicit constructions of randomness extractors (see Appendix B or [62]). In this case, the predicate f is viewed as (an error correcting encoding of) a somewhat random function, and the construction makes sense because it refers to f in a black-box manner. In the analysis we rely on the fact that f can be approximated by combining relatively little information (regarding f) with (black-box access to) a distinguisher for G^f. For further details see Section B.2.

3.3 Reflections Regarding Derandomization

Part 1 of Theorem 3.6 is often summarized by saying that (under some reasonable assumptions) *randomness is useless*. We believe that this interpretation is wrong even within the restricted context of traditional complexity classes, and is bluntly wrong if taken outside of the latter context. Let us elaborate.

Taking a closer look at the proof of Theorem 3.3 (which underlies Theorem 3.6), we note that a randomized algorithm A of time-complexity t is emulated by a deterministic algorithm A' of time complexity $t' = \text{poly}(t)$. Further noting that $A' = A_G$ invokes A (as well as the canonical derandomizer G) for $\Omega(t)$ times (because $\ell(k) = O(2^k)$ implies $2^k = \Omega(t)$), we infer that $t' = \Omega(t^2)$ must hold. Thus, derandomization via (Part 1 of) Theorem 3.6 is not really for free.

More importantly, we note that derandomization is not possible in various distributed settings, when both parties may protect their conflicting interests by employing randomization. Notable examples include most cryptographic primitives (e.g., encryption) as well as most types of probabilistic proof systems (e.g., PCP). Additional settings where randomness makes a difference (either between impossibility and possibility or between formidable and affordable cost) include distributed computing (see [8]), communication complexity (see [39]), parallel architectures (see [40]), sampling (see, e.g., [24, Apdx. D.3]), and property testing (see, e.g., [24, Sec. 10.1.2]).

Notes

As observed by Yao [73], a non-uniformly strong notion of pseudorandom generators yields non-trivial derandomization of time-complexity classes. A key observation of Nisan [49, 52] is that whenever a pseudorandom generator is used in this way, it suffices to require that the generator runs in time that is exponential in its seed length, and so the generator may have running-time greater than the distinguisher (representing the algorithm to be derandomized). This observation motivates the definition of canonical derandomizers as well as the construction of Nisan and Wigderson [49, 52], which is the basis for further improvements culminating in [34]. Part 1 of Theorem 3.6 (i.e., the so-called "high end" derandomization of $\mathcal{BPP}$) is due to Impagliazzo and Wigderson [34], whereas Part 2 (the "low end") is from [52].

The Nisan–Wigderson Generator [52] was subsequently used in several ways transcending its original presentation. We mention its application towards fooling nondeterministic machines (and thus derandomizing constant-round interactive proof systems) and to the construction of randomness extractors (see [65] as well as [62]).

In contrast to the aforementioned derandomization results, which place $\mathcal{BPP}$ in some worst-case deterministic complexity class based on some non-uniform (worst-case) assumption, we now mention a result that places $\mathcal{BPP}$ in an average-case deterministic complexity class based on a uniform-complexity (worst-case) assumption. We refer specifically to a theorem, which is due to Impagliazzo and Wigderson [35] (but is not presented in the main text), that asserts the following: *if $\mathcal{BPP}$ is not contained in $\mathcal{EXP}$* (almost-everywhere) *then $\mathcal{BPP}$ has deterministic subexponential time algorithms that are correct on all typical cases* (i.e., with respect to any polynomial-time sampleable distribution).

In Section 3.2.3 we mentioned that Construction 3.4, instantiated with the `parity` function, yields a pseudorandom generator that fools $\mathcal{AC}^0$ while using a seed of polylogarithmic length. Alternative constructions follow by a recent result of [12] that asserts that polylogarithmic-wise independence generators (see, e.g., Proposition 5.1) fool $\mathcal{AC}^0$.

Exercises

Exercise 3.1 Show that Construction 2.7 may fail in the context of canonical derandomizers. Specifically, prove that it fails for the canonical derandomizer G' that is presented in the proof of Theorem 3.5.

Exercise 3.2 In relation to Definition 3.1 (and assuming $\ell(k) > k$), show that there exists a circuit of size $O(2^k \cdot \ell(k))$ that violates Eq. (3.1).

Guideline: The circuit may incorporate all values in the range of G and decide by comparing its input to these values.

Exercise 3.3 (constructing a set system for Theorem 3.5) For every $\gamma > 0$, show a construction of a set system S as in Condition 2 of Theorem 3.5, with $m(k) = \Omega(k)$ and $\ell(k) = 2^{\Omega(k)}$.

Guideline: We assume, without loss of generality, that $\gamma < 1$, and set $m(k) = (\gamma/2) \cdot k$ and $\ell(k) = 2^{\gamma m(k)/6}$. We construct the set system $S_1, ..., S_{\ell(k)}$ in iterations, selecting S_i as the first $m(k)$-subset of $[k]$ that has sufficiently small intersections with each of the previous sets $S_1, ..., S_{i-1}$. The existence of such a set S_i can be proved using the Probabilistic Method (cf. [6]). Specifically, for a fixed $m(k)$-subset S', the probability that a random $m(k)$-subset has intersection greater than $\gamma m(k)$ with S' is smaller than $2^{-\gamma m(k)/6}$, because the expected intersection size is $(\gamma/2) \cdot m(k)$. Thus, with positive probability a random $m(k)$-subset has intersection of size at most $\gamma m(k)$ with each of the previous $i-1 < \ell(k) = 2^{\gamma m(k)/6}$ subsets. Note that we construct S_i in time $\binom{k}{m(k)} \cdot (i-1) \cdot m(k) < 2^k \cdot \ell(k) \cdot k$, and thus S is computable in time $k2^k \cdot \ell(k)^2 < 2^{2k}$.

Exercise 3.4 (pseudorandom vs. unpredictability, by circuits) In continuation to Exercise 2.11, show that if there exists a circuit of size s that distinguishes Z_n from U_ℓ with gap δ, then there exists an $i < \ell = |Z_n|$ and a circuit of size $s + O(1)$

that given an i-bit long prefix of Z_n guesses the $i+1^{\text{st}}$ bit with success probability at least $\frac{1}{2}+\frac{\delta}{\ell}$.

Guideline: Defining hybrids as in Exercise 2.11, note that, for some i, the given circuit distinguishes the ith hybrid from the $i+1^{\text{st}}$ hybrid with gap at least δ/ℓ.

Exercise 3.5 Suppose that the sets S_i's in Construction 3.4 are disjoint and that $f:\{0,1\}^m \to \{0,1\}$ is T-inapproximable. Prove that for every circuit C of size $T-O(1)$ it holds that $|\mathsf{Pr}[C(G(U_k))=1]-\mathsf{Pr}[C(U_\ell)=1]| < \ell/T$.

Guideline: Prove the contrapositive using Exercise 3.4. Note that the value of the $i+1^{\text{st}}$ bit of $G(U_k)$ is statistically independent of the values of the first i bits of $G(U_k)$, and thus predicting it yields an approximator for f. Indeed, such an approximator can be obtained by fixing the first i bits of $G(U_k)$ via an averaging argument.

Exercise 3.6 (Theorem 3.5, generalized) Let $\ell, m, m', T:\mathbb{N}\to\mathbb{N}$ satisfy $\ell(k)^2 + \widetilde{O}(\ell(k)2^{m'(k)}) < T(m(k))$. Suppose that the following two conditions hold:

1. There exists an exponential-time computable function $f:\{0,1\}^* \to \{0,1\}$ that is T-inapproximable.
2. There exists an exponential-time computable function $S:\mathbb{N}\times\mathbb{N}\to 2^{\mathbb{N}}$ such that for every k and $i=1,...,\ell(k)$ it holds that $S(k,i)\subseteq [k]$ and $|S(k,i)|=m(k)$, and $|S(k,i)\cap S(k,j)| \le m'(k)$ for every k and $i \ne j$.

Prove that using G as defined in Construction 3.4, with $S_i = S(k,i)$, yields a canonical derandomizer with stretch ℓ.

Guideline: Following the proof of Theorem 3.5, just note that the circuit constructed for approximating $f(U_{m(k)})$ has size $\ell(k)^2 + \ell(k)\cdot\widetilde{O}(2^{m'(k)})$ and success probability at least $(1/2)+(1/7\ell(k))$.

Exercise 3.7 (Part 2 of Theorem 3.6) Prove that if for every polynomial T there exists a T-inapproximable predicate in $\mathcal{E}$, then $\mathcal{BPP} \subseteq \bigcap_{\varepsilon>0} \mathrm{DTIME}(t_\varepsilon)$, where $t_\varepsilon(n) \stackrel{\text{def}}{=} 2^{n^\varepsilon}$.

Guideline: Using Proposition 3.2, it suffices to present, for every polynomial p and every constant $\varepsilon>0$, a canonical derandomizer of stretch $\ell(k)=p(k^{1/\varepsilon})$. Such a derandomizer can be presented by applying Exercise 3.6 using $m(k)=\sqrt{k}$, $m'(k)=O(\log k)$, and $T(m(k)) = \ell(k)^2+\widetilde{O}(\ell(k)2^{m'(k)})$. Note that T is a polynomial, revisit Exercise 3.3 in order to obtain a set system as required in Exercise 3.6 (for these parameters), and use [24, Thm. 7.10].

Exercise 3.8 (canonical derandomizers imply hard problems) Prove that the hardness hypothesis made in each part of Theorem 3.6 is essential for the existence of a corresponding canonical derandomizer. More generally, prove that the existence of a canonical derandomizer with stretch ℓ implies the existence of a predicate in $\mathcal{E}$ that is T-inapproximable for $T(n)=\ell(n)^{1/O(1)}$.

Guideline: We focus on obtaining a predicate in $\mathcal{E}$ that cannot be computed by circuits of size ℓ, and note that the claim follows by applying the techniques in [24, §7.2.1.3]. Given a canonical derandomizer $G:\{0,1\}^k \to \{0,1\}^{\ell(k)}$, we consider the predicate $f:\{0,1\}^{k+1} \to \{0,1\}$ that satisfies $f(x)=1$ if and only if there exists $s\in\{0,1\}^{|x|-1}$ such that x is a prefix of $G(s)$. Note that f is in $\mathcal{E}$ and that an algorithm computing f yields a distinguisher of $G(U_k)$ and $U_{\ell(k)}$.

Chapter 4

Space-Bounded Distinguishers

In the previous two chapters we have considered generators that output sequences that look random to any efficient procedure, where the latter were modeled by time-bounded computations. Specifically, in Chapter 2 we considered indistinguishability by polynomial-time procedures. A finer classification of time-bounded procedures is obtained by considering their *space-complexity*; that is, restricting the space-complexity of time-bounded computations. This restriction leads to the notion of pseudorandom generators that fool space-bounded distinguishers. Interestingly, in contrast to the notions of pseudorandom generators that were considered in Chapters 2 and 3, the existence of pseudorandom generators that fool space-bounded distinguishers can be established without relying on computational assumptions.

Prerequisites: Technically speaking, the current chapter is self-contained, but various definitional choices are justified by reference to the standard definitions of space-bounded randomized algorithms. Thus, a review of that model (as provided in, e.g., [24, Sec. 6.1.5]) is recommended as conceptual background for the current chapter.

4.1 Definitional Issues

Our main motivation for considering space-bounded distinguishers is to develop a notion of pseudorandomness that is adequate for space-bounded randomized algorithms. That is, such algorithms should essentially maintain their behavior when their source of internal coin tosses is replaced by a source of pseudorandom bits (which may be generated based on a much shorter random seed). We thus start by recalling and reviewing the natural notion of space-bounded randomized algorithms.

Unfortunately, natural notions of space-bounded computations are quite subtle, especially when non-determinism or randomization are concerned (see [24, Sec. 5.3] and [24, Sec. 6.1.5], respectively). Two major definitional issues regarding randomized space-bounded computations are the need for imposing explicit *time bounds* and the type of *access to the random tape*.

1. Time bounds: The question is whether or not the space-bounded machines are restricted to time-complexity that is at most exponential in their space-complexity.[1] Recall that such an upper-bound follows automatically in the deterministic case, and can be assumed (without loss of generality) in the non-deterministic case, *but it does not necessarily hold in the randomized case.* Furthermore, failing to restrict the time-complexity of randomized space-bounded machines makes them unnatural and unintentionally too strong (e.g., capable of emulating non-deterministic computations with no overhead in terms of space-complexity).

   Seeking a natural model of randomized space-bounded algorithms, we postulate that their time-complexity must be at most exponential in their space-complexity.

2. Access to the random tape: Recall that randomized algorithms may be modeled as machines that are provided with the necessary randomness via a special random-tape. The question is whether the space-bounded machine has uni-directional or bi-directional (i.e., unrestricted) access to its random-tape. (Allowing bi-directional access means that the randomness is recorded "for free"; that is, without being accounted for in the space-bound.)

   Recall that uni-directional access to the random-tape corresponds to the natural model of an on-line randomized machine, which determines its moves based on its internal coin tosses (and thus cannot record its past coin tosses "for free"). Thus, we consider uni-directional access.[2]

Hence, we focus on randomized space-bounded computations that have time-complexity that is at most exponential in their space-complexity and access their random-tape in a uni-directional manner.

When seeking a notion of pseudorandomness that is adequate for the foregoing notion of randomized space-bounded computations, we note that the corresponding distinguisher is obtained by fixing the main input of the computation and *viewing the contents of the random-tape of the computation as the only input of the distinguisher.* Thus, in accordance with the foregoing notion of randomized space-bounded computation, *we consider space-bounded distinguishers that have a uni-directional access to the input sequence that they examine.* Let us consider the type of algorithms that arise.

We consider *space-bounded algorithms that have a uni-directional access to their input.* At each step, based on the contents of its temporary storage, such an algorithm may either read the next input bit or stay at the current location on the input, where in either case the algorithm may modify its temporary storage. To simplify our analysis of such algorithms, we consider a corresponding *non-uniform model* in which, at each step, the algorithm reads the next input bit and updates its temporary

---

[1] Alternatively, one can ask whether these machines must always halt or only halt with probability approaching 1. It can be shown that the only way to ensure "absolute halting" is to have time-complexity that is at most exponential in the space-complexity. (In the current discussion as well as throughout this chapter, we assume that the space-complexity is at least logarithmic.)

[2] We note that the fact that we restrict our attention to uni-directional access is instrumental in obtaining space-robust generators without making intractability assumptions. Analogous generators for bi-directional space-bounded computations would imply hardness results of a breakthrough nature in the area.

storage according to an arbitrary function applied to the previous contents of that storage (and to the new bit). Note that we have strengthened the model by allowing arbitrary (updating) functions, which can be implemented by (non-uniform) circuits having size that is exponential in the space-bound, rather than using (updating) functions that can be (uniformly) computed in time that is exponential in the space-bound. This strengthening is motivated by the fact that the known constructions of pseudorandom generators remain valid also when the space-bounded distinguishers are non-uniform and by the fact that non-uniform distinguishers arise anyhow in derandomization.

The computation of the foregoing non-uniform space-bounded algorithms (or automata)[3] can be represented by directed layered graphs, where the vertices in each layer correspond to possible contents of the temporary storage and transition between neighboring layers corresponds to a step of the computation. Foreseeing the application of this model for the description of potential distinguishers, we parameterize these layered graphs based on the index, denoted k, of the relevant ensembles (e.g., $\{G(U_k)\}_{k\in\mathbb{N}}$ and $\{U_{\ell(k)}\}_{k\in\mathbb{N}}$). That is, we present both the input length, denoted $\ell = \ell(k)$, and the space-bound, denoted $s(k)$, as functions of the parameter k. Thus, we define a non-uniform automaton of space $s:\mathbb{N}\rightarrow\mathbb{N}$ (and depth $\ell:\mathbb{N}\rightarrow\mathbb{N}$) as a family, $\{D_k\}_{k\in\mathbb{N}}$, of directed layered graphs with labeled edges such that the following conditions hold:

- The digraph D_k consists of $\ell(k)+1$ layers, each containing at most $2^{s(k)}$ vertices. The first layer contains a single vertex, which is the digraph's (single) source (i.e., a vertex with no incoming edges), and the last layer contains all the digraph's sinks (i.e., vertices with no outgoing edges).

- The only directed edges in D_k are between adjacent layers, going from layer i to layer $i+1$, for $i \leq \ell(k)$. These edges are labeled such that each (non-sink) vertex of D_k has two (possibly parallel) outgoing directed edges, one labeled 0 and the other labeled 1.

The result of the computation of such an automaton, on an input of adequate length (i.e., length ℓ where D_k has $\ell+1$ layers), is defined as the vertex (in last layer) reached when following the sequence of edges that are labeled by the corresponding bits of the input. That is, on input $x = x_1 \cdots x_\ell$, in the ith step (for $i = 1, ..., \ell$) we move from the current vertex (which resides in the ith layer) to one of its neighbors (which resides in the $i+1^{\text{st}}$ layer) by following the outgoing edge labeled x_i. Using a fixed partition of the vertices of the last layer, this defines a natural notion of a decision (by D_k); that is, we write $D_k(x) = 1$ if on input x the automaton D_k reached a vertex that belongs to the first part of the aforementioned partition.

Definition 4.1 (indistinguishability by space-bounded automata):

[3]We use the term automaton (rather than algorithm or machine) in order to remind the reader that this computing device reads its input in a uni-directional manner. Alternative terms that may be used are "real-time" or "on-line" machines. We prefer not using the term "on-line" machine in order to keep a clear distinction between our notion and randomized algorithms that have free access to their input (and on-line access to a source of randomness). Indeed, the automata considered here arise from the latter algorithms by fixing their primary input and considering the random source as their (only) input. We also note that the automata considered here are a special case of Ordered Binary Decision Diagrams (OBDDs; see [71]).

- *For a non-uniform automaton,* $\{D_k\}_{k\in\mathbb{N}}$*, and two probability ensembles,* $\{X_k\}_{k\in\mathbb{N}}$ *and* $\{Y_k\}_{k\in\mathbb{N}}$*, the function* $d:\mathbb{N}\to[0,1]$ *defined as*

$$d(k) \stackrel{\text{def}}{=} |\mathsf{Pr}[D_k(X_k)=1] - \mathsf{Pr}[D_k(Y_k)=1]|$$

is called the distinguishability-gap *of* $\{D_k\}$ *between the two ensembles.*

- *Let* $s:\mathbb{N}\to\mathbb{N}$ *and* $\varepsilon:\mathbb{N}\to[0,1]$*. A probability ensemble,* $\{X_k\}_{k\in\mathbb{N}}$*, is called* (s,ε)*-*pseudorandom *if for any non-uniform automaton of space* $s(\cdot)$*, the distinguishability-gap of the automaton between* $\{X_k\}_{k\in\mathbb{N}}$ *and the corresponding uniform ensemble (i.e.,* $\{U_{|X_k|}\}_{k\in\mathbb{N}}$*) is at most* $\varepsilon(\cdot)$*.*

- *A deterministic algorithm* G *of stretch function* ℓ *is called an* (s,ε)-pseudorandom generator *if the ensemble* $\{G(U_k)\}_{k\in\mathbb{N}}$ *is* (s,ε)*-pseudorandom. That is, every non-uniform automaton of space* $s(\cdot)$ *has a distinguishing gap of at most* $\varepsilon(\cdot)$ *between* $\{G(U_k)\}_{k\in\mathbb{N}}$ *and* $\{U_{\ell(k)}\}_{k\in\mathbb{N}}$*.*

Thus, when using a random seed of length k, an (s,ε)-pseudorandom generator outputs a sequence of length $\ell(k)$ that looks random to observers having space $s(k)$. Note that $s(k)\le k$ is a necessary condition for the existence of $(s,0.5)$-pseudorandom generators, because a non-uniform automaton of space $s(k)>k$ can recognize the image of a generator (which contains at most 2^k strings of length $\ell(k)>k$). More generally, there is a trade-off between $k-s(k)$ and the stretch of (s,ε)-pseudorandom generators; for details see Exercises 4.1 and 4.2.

Note: We stated the space-bound of the potential distinguisher (as well as the stretch function) in terms of the seed-length, denoted k, of the generator. In contrast, other sources present a parameterization in terms of the space-bound of the potential distinguisher, denoted m. The translation is obtained by using $m=s(k)$, and we shall provide it subsequent to the main statements of Theorems 4.2 and 4.3.

4.2 Two Constructions

In contrast to the case of pseudorandom generators that fool time-bounded distinguishers, pseudorandom generators that fool space-bounded distinguishers can be constructed without relying on any computational assumption. The following two theorems exhibit two rather extreme cases of a general trade-off between the space-bound of the potential distinguisher and the stretch function of the generator.[4] We stress that both theorems fall short of providing parameters as in Exercise 4.2, but they refer to relatively efficient constructions. We start with an attempt to maximize the stretch.

Theorem 4.2 (stretch exponential in the space-bound for $s(k)=\sqrt{k}$): *For every space constructible function* $s:\mathbb{N}\to\mathbb{N}$*, there exists an* $(s,2^{-s})$*-pseudorandom generator of stretch function* $\ell(k)=\min(2^{k/O(s(k))},2^{s(k)})$*. Furthermore, the generator works in space that is linear in the length of the seed, and in time that is linear in the stretch function.*

[4]These two results have been "interpolated" in [7]: There exists a parameterized family of "space fooling" pseudorandom generators that includes both results as extreme special cases.

In other words, for every $t \leq m$, we have a generator that takes a random seed of length $k = O(t \cdot m)$ and produces a sequence of length 2^t that looks random to any (non-uniform) automaton of space m (up to a distinguishing gap of 2^{-m}). In particular, using a random seed of length $k = O(m^2)$, one can produce a sequence of length 2^m that looks random to any (non-uniform) automaton of space m. Thus, *one may replace random sequences used by any space-bounded computation, by sequences that are efficiently generated from random seeds of length quadratic in the space bound.* The common instantiation of the latter assertion is for log-space algorithms. In Section 4.2.2, we apply Theorem 4.2 (and its underlying ideas) for the derandomization of space-complexity classes such as $\mathcal{BPL}$ (i.e., the log-space analogue of $\mathcal{BPP}$). Theorem 4.2 itself is proved in Section 4.2.1.

We now turn to the case where one wishes to maximize the space-bound of potential distinguishers. We warn that Theorem 4.3 only guarantees a subexponential distinguishing gap (rather than the exponential distinguishing gap guaranteed in Theorem 4.2).

Theorem 4.3 (polynomial stretch and linear space-bound): *For any polynomial p and for some $s(k) = k/O(1)$, there exists an $(s, 2^{-\sqrt{s}})$-pseudorandom generator of stretch function p. Furthermore, the generator works in linear-space and polynomial-time* (both stated in terms of the length of the seed).

In other words, we have a generator that takes a random seed of length $k = O(m)$ and produces a sequence of length $\text{poly}(m)$ that looks random to any (non-uniform) automaton of space m. Thus, one may *convert any randomized computation utilizing polynomial-time and linear-space into a functionally equivalent randomized computation of similar time and space complexities that uses only a linear number of coin tosses.*

4.2.1 Sketches of the proofs of Theorems 4.2 and 4.3

In both cases, we start the proof by considering a generic space-bounded distinguisher and show that the input distribution that this distinguisher examines can be modified (from the uniform distribution into a pseudorandom one) without having the distinguisher notice the difference. This modification (or rather a sequence of modifications) yields a construction of a pseudorandom generator, which is only spelled out at the end of the argument.

Sketch of the proof of Theorem 4.2 (see details in [50])

The main technical tool used in this proof is the "mixing property" of pairwise independent hash functions (see Appendix A). A family of functions H_n, which map $\{0,1\}^n$ to itself, is called mixing if for every pair of subsets $A, B \subseteq \{0,1\}^n$ for all but very few (i.e., $\exp(-\Omega(n))$ fraction) of the functions $h \in H_n$, it holds that

$$\mathsf{Pr}[U_n \in A \wedge h(U_n) \in B] \approx \frac{|A|}{2^n} \cdot \frac{|B|}{2^n} \tag{4.1}$$

where the approximation is up to an additive term of $\exp(-\Omega(n))$. (See the generalization of Lemma A.4, which implies that $\exp(-\Omega(n))$ can be set to $2^{-n/3}$.)

We may assume, without loss of generality, that $s(k) = \Omega(\sqrt{k})$, and thus $\ell(k) \leq 2^{s(k)}$ holds. For any $s(k)$-space distinguisher D_k as in Definition 4.1, we consider an auxiliary "distinguisher" D'_k that is obtained by "contracting" every block of $n \stackrel{\text{def}}{=} \Theta(s(k))$ consecutive layers in D_k, yielding a directed layered graph with $\ell' \stackrel{\text{def}}{=} \ell(k)/n < 2^{s(k)}$ layers (and $2^{s(k)}$ vertices in each layer). Specifically,

- each vertex in D'_k has 2^n (possibly parallel) directed edges going to various vertices of the next level; and
- each such edge is labeled by an n-bit long string such that the directed edge (u, v) labeled $\sigma_1\sigma_2 \cdots \sigma_n$ in D'_k replaces the n-edge directed path between u and v in D_k that consists of edges labeled $\sigma_1, \sigma_2, ..., \sigma_n$.

The graph D'_k simulates D_k in the obvious manner; that is, the computation of D'_k on an input of length $\ell(k) = \ell' \cdot n$ is defined by breaking the input into consecutive substrings of length n and following the path of edges that are labeled by the corresponding n-bit long substrings.

The key observation is that D'_k cannot distinguish between a random $\ell' \cdot n$-bit long input (i.e., $U_{\ell' \cdot n} \equiv U_n^{(1)} U_n^{(2)} \cdots U_n^{(\ell')}$) and a "pseudorandom" input of the form $U_n^{(1)} h(U_n^{(1)}) U_n^{(2)} h(U_n^{(2)}) \cdots U_n^{(\ell'/2)} h(U_n^{(\ell'/2)})$, where $h \in H_n$ is a (suitably fixed) hash function. To prove this claim, we consider an arbitrary pair of neighboring vertices, u and v (in layers i and $i+1$, respectively), and denote by $L_{u,v} \subseteq \{0,1\}^n$ the set of the labels of the edges going from u to v. Similarly, for a vertex w at layer $i+2$, we let $L'_{v,w}$ denote the set of the labels of the edges going from v to w. By Eq. (4.1), for all but very few of the functions $h \in H_n$, it holds that

$$\mathsf{Pr}[U_n \in L_{u,v} \wedge h(U_n) \in L'_{v,w}] \approx \mathsf{Pr}[U_n \in L_{u,v}] \cdot \mathsf{Pr}[U_n \in L'_{v,w}] \,, \tag{4.2}$$

where "very few" and $\approx$ are as in Eq. (4.1). Thus, for all but $\exp(-\Omega(n))$ fraction of the choices of $h \in H_n$, *replacing the coins in the second transition* (i.e., the transition from layer $i+1$ to layer $i+2$) *with the value of* h *applied to the outcomes of the coins used in the first transition* (i.e., the transition from layer i to $i+1$), *approximately maintains the probability that* D'_k *moves from* u *to* w *via* v. Using a union bound (on all triples (u, v, w) as in the foregoing), we note that, for all but $2^{3s(k)} \cdot \ell' \cdot \exp(-\Omega(n))$ fraction of the choices of $h \in H_n$, the foregoing replacement approximately maintains the probability that D'_k moves through any specific two-edge path of D'_k.

Using $\ell' < 2^{s(k)}$ and a suitable choice of $n = \Theta(s(k))$, it holds that $2^{3s(k)} \cdot \ell' \cdot \exp(-\Omega(n)) < \exp(-\Omega(n))$, and thus all but a "few" functions $h \in H_n$ are good for approximating all of these transition probabilities. (We stress that the same h can be used in all of these approximations.) Thus, *at the cost of extra* $|h|$ *random bits, we can reduce the number of true random coins used in transitions on* D'_k *by a factor of two, without significantly affecting the final decision of* D'_k (where again we use the fact that $\ell' \cdot \exp(-\Omega(n)) < \exp(-\Omega(n))$, which implies that the approximation errors do not accumulate to too much). In other words, at the cost of extra $|h|$ random bits, we can effectively contract the distinguisher to half its length while approximately maintaining the probability that the distinguisher accepts a random input. That is, fixing a good h (i.e., one that provides a good approximation to the transition probability over all $2^{3s(k)} \cdot \ell'$ two-edge paths), we can replace the two-edge paths in D'_k by edges in a new distinguisher D''_k (which depends on h) such that an edge

(u,w) labeled $r \in \{0,1\}^n$ appears in D''_k if and only if, for some v, the path (u,v,w) appears in D'_k with the first edge (i.e., (u,v)) labeled r and the second edge (i.e., (v,w)) labeled $h(r)$. Needless to say, the crucial point is that $\Pr[D''_k(U_{(\ell'/2)\cdot n})=1]$ approximates $\Pr[D'_k(U_{\ell'\cdot n})=1]$.

The foregoing process can be applied to D''_k resulting in a distinguisher D'''_k of half the length, and so on. Each time we contract the current distinguisher by a factor of two, and do so by randomly selecting (and fixing) a new hash function. Thus, repeating the process for a logarithmic (in the depth of D'_k) number of times we obtain a distinguisher that only examines n bits, at which point we stop. In total, we have used $t \stackrel{\text{def}}{=} \log_2(\ell'/n) < \log_2 \ell(k)$ random hash functions. This means that we can generate a (pseudorandom) sequence that fools the original D_k by using a seed of length $n + t \cdot \log_2 |H_n|$. Using $n = \Theta(s(k))$ and an adequate family H_n (which, in particular, satisfies $|H_n| = 2^{O(n)}$), we obtain the desired $(s, 2^{-s})$-pseudorandom generator, which indeed uses a seed of length $O(s(k) \cdot \log_2 \ell(k)) = k$.

Digest. The actual proof of Theorem 4.4 refers to a stronger class of distinguishers that read n-bit long blocks at a time, and process each such block arbitrarily (as long as the space occupied before and after reading this block is upper-bounded by $s(n)$).[5] Thus, the foregoing pseudorandom generator fools this stronger type of distinguishers, which was used in order to facilitate the argument.

Rough sketch of the proof of Theorem 4.3 (see details in [53])

The main technical tool used in this proof is a suitable randomness extractor (see Appendix B), which is indeed a much more powerful tool than hashing functions. The basic idea is that when the distinguisher D_k is at some "distant" layer, say at layer $t = \Omega(s(k))$, it typically "knows" little about the random choices that led it there. That is, D_k has only $s(k)$ bits of memory, which leaves out $t - s(k)$ bits of "uncertainty" (or randomness) regarding the previous moves. Thus, much of the randomness that led D_k to its current state may be "reused" (or "recycled"). To reuse these bits we need to extract *almost* uniform distribution on strings of sufficient length out of the aforementioned distribution (over $\{0,1\}^t$) that has entropy[6] at least $t - s(k)$. Furthermore, such an extraction requires some additional truly random bits, yet relatively few such bits. In particular, using $k' = \Omega(\log t)$ bits towards this end, the extracted bits are $\exp(-\Omega(k'))$ away from uniform.

The gain from the aforementioned recycling is significant if recycling is repeated sufficiently many times. Towards this end, we break the k-bit long seed into two parts, denoted $r' \in \{0,1\}^{k/2}$ and $(r_1, ..., r_{3\sqrt{k}})$, where $|r_i| = \sqrt{k}/6$, and set $n = k/3$. Intuitively, r' will be used for determining the first n steps, and it will be reused (or recycled) together with r_i for determining the steps $i \cdot n + 1$ through $(i+1) \cdot n$. Looking at layer $i \cdot n$, we consider the information regarding r' that is "known" to D_k (when reaching a specific vertex at layer $i \cdot n$). Typically, the conditional distribution of r', given that we reached a specific vertex at layer $i \cdot n$, has (min-)entropy greater than $0.99 \cdot ((k/2) - s(k))$. Using r_i (as a seed of an extractor applied to r'), we can extract

[5] This extra distinguishing power is referred to in [66, Sec. 3.4.2].

[6] Actually, a stronger technical condition needs to be and can be imposed on the latter distribution. Specifically, with overwhelmingly high probability, at layer t, automaton D_k is at a vertex that can be reached in more than $2^{0.99 \cdot (t - s(k))}$ different ways. In this case, the distribution representing a random walk that reaches this vertex has min-entropy greater than $0.99 \cdot (t - s(k))$.

$0.9\cdot((k/2)-s(k)-o(k)) > k/3 = n$ bits that are almost-random (i.e., $2^{-\Omega(\sqrt{k})}$-close to U_n) with respect to D_k, and use these bits for determining the next n steps. Hence, using k random bits, we produce a sequence of length $(1+3\sqrt{k})\cdot n > k^{3/2}$ that fools automata of space bound, say, $s(k) = k/10$. Specifically, using an extractor of the form $\text{Ext} : \{0,1\}^{k/2} \times \{0,1\}^{\sqrt{k}/6} \to \{0,1\}^{k/3}$, we map the seed $(r', r_1, ..., r_{3\sqrt{k}})$ to the output sequence $(r', \text{Ext}(r', r_1), ..., \text{Ext}(r', r_{3\sqrt{k}}))$. Thus, *we obtained an* $(s, 2^{-\Omega(\sqrt{s})})$*-pseudorandom generator of stretch function* $\ell(k) = k^{3/2}$.

In order to obtain an arbitrary polynomial stretch rather than a specific polynomial stretch (i.e., $\ell(k) = k^{3/2}$), we iteratively compose generators as above with themselves (for a constant number of times). The basic composition combines an (s_1, ε_1)-pseudorandom generator of stretch function ℓ_1, denoted G_1, with an (s_2, ε_2)-pseudorandom generator of stretch function ℓ_2, denoted G_2. On input $s \in \{0,1\}^k$, the resulting generator first computes $G_1(s)$, parses $G_1(s)$ into t consecutive k'-bit long blocks, where $k' = s_1(k)/2$ and $t = \ell_1(k)/k'$, and applies G_2 to each block (outputting the concatenation of the t results). This generator, denoted G, has stretch $\ell(k) = t \cdot \ell_2(k')$, and for $s_1(k) = \Theta(k)$ we have $\ell(k) = \ell_1(k) \cdot \ell_2(\Omega(k))/O(k)$. The pseudorandomness of G can be established via a hybrid argument (which refers to the intermediate hybrid distribution $G_2(U_{k'}^{(1)}) \cdots G_2(U_{k'}^{(t)})$ and uses the fact that the second step in the computation of G can be performed by a non-uniform automaton of space $s_1/2$).

4.2.2 Derandomization of space-complexity classes

As a direct application of Theorem 4.2, we obtain that $\mathcal{BPL} \subseteq \text{DSPACE}(\log^2)$, where $\mathcal{BPL}$ denotes the log-space analogue of $\mathcal{BPP}$. (Recall that $\mathcal{NL} \subseteq \text{DSPACE}(\log^2)$, but it is not known whether or not $\mathcal{BPL} \subseteq \mathcal{NL}$.)[7] A stronger derandomization result can be obtained by a finer analysis of the proof of Theorem 4.2.

Theorem 4.4 *$\mathcal{BPL} \subseteq \mathcal{SC}$, where $\mathcal{SC}$ denotes the class of decision problems that can be solved by deterministic algorithms that run in polynomial-time and polylogarithmic-space.*

Thus, $\mathcal{BPL}$ (and, in particular, $\mathcal{RL} \subseteq \mathcal{BPL}$) is placed in a class not known to contain $\mathcal{NL}$. Another such result was subsequently obtained in [59]: Randomized log-space can be simulated in deterministic space $o(\log^2)$; specifically, in space $\log^{3/2}$. We mention that the archetypical problem of $\mathcal{RL}$ was recently proved to be in $\mathcal{L}$ (see [56]).

Sketch of the proof of Theorem 4.4 (see details in [51])

We are going to use the generator construction provided in the proof of Theorem 4.2, but we will show that the main part of the seed (i.e., the sequence of hash functions) can be fixed (depending on the distinguisher at hand). Furthermore, this fixing can be performed in polylogarithmic space and polynomial-time. Specifically, wishing to derandomize a specific log-space computation (which refers to a specific input), we first obtain the corresponding distinguisher, denoted D'_k, that represents this

[7] Indeed, the log-space analogue of $\mathcal{RP}$, denoted $\mathcal{RL}$, is contained in $\mathcal{NL} \subseteq \text{DSPACE}(\log^2)$, and thus the fact that Theorem 4.2 implies $\mathcal{RL} \subseteq \text{DSPACE}(\log^2)$ is of no interest.

computation (as a function of the outcomes of the internal coin tosses of the log-space algorithm). The key observation is that the question of whether or not a specific hash function $h \in H_n$ is good for a specific D'_k can be determined in space that is linear in $n = |h|/2$ and logarithmic in the size of D'_k. Indeed, the time-complexity of this decision procedure is exponential in its space-complexity. It follows that we can find a good $h \in H_n$, for a given D'_k, within these complexities (by scanning through all possible $h \in H_n$). Once a good h is found, we can also construct the corresponding graph D''_k (in which edges represent two-edge paths in D'_k), again within the same complexity. Actually, it will be more instructive to note that we can determine a step (i.e., an edge-traversal) in D''_k by making two steps (edge-traversals) in D'_k. This will allow us to fix a hash function for D''_k, and so on. Details follow.

The main claim is that the entire process of finding a sequence of $t \stackrel{\text{def}}{=} \log_2 \ell'(k)$ good hash functions can be performed in space $t \cdot O(n+\log|D_k|) = O(n+\log|D_k|)^2$ and time $\text{poly}(2^n \cdot |D_k|)$; that is, the time-complexity is sub-exponential in the space-complexity (i.e., the time-complexity is significantly smaller than the generic bound of $\exp(O(n + \log|D_k|)^2))$. Starting with $D_k^{(1)} = D'_k$, we find a good (for $D_k^{(1)}$) hashing function $h^{(1)} \in H_n$, which defines $D_k^{(2)} = D''_k$. Having found (and stored) $h^{(1)}, ..., h^{(i)} \in H_n$, which determine $D_k^{(i+1)}$, we find a good hashing function $h^{(i+1)} \in H_n$ for $D_k^{(i+1)}$ by emulating pairs of edge-traversals on $D_k^{(i+1)}$. Indeed, a key point is that we do *not* construct the sequence of graphs $D_k^{(2)}, ..., D_k^{(i+1)}$, but rather emulate an edge-traversal in $D_k^{(i+1)}$ by making 2^i edge-traversals in D'_k, using $h^{(1)}, ..., h^{(i)}$: The (edge-traversal) move $\alpha \in \{0,1\}^n$ starting at vertex v of $D_k^{(i+1)}$ translates to a sequence of 2^i moves starting at vertex v of D'_k, where the moves are determined by the 2^i-long sequence (of n-bit strings)

$$\overline{h}^{(0^i)}(\alpha), \overline{h}^{(0^{i-2}01)}(\alpha), \overline{h}^{(0^{i-2}10)}(\alpha), \overline{h}^{(0^{i-2}11)}(\alpha), ..., \overline{h}^{(1^i)}(\alpha),$$

where $\overline{h}^{(\sigma_i \cdots \sigma_1)}$ is the function obtained by the composition of a subsequence of the functions $h^{(i)}, ..., h^{(1)}$ determined by $\sigma_i \cdots \sigma_1$. Specifically, $\overline{h}^{(\sigma_i \cdots \sigma_1)}$ equals $h^{(i_{t'})} \circ \cdots \circ h^{(i_2)} \circ h^{(i_1)}$, where $i_1 < i_2 < \cdots < i_{t'}$ and $\{i_j : j{=}1, ..., t'\} = \{j : \sigma_j{=}1\}$.

Recall that the ability to perform edge-traversals on $D_k^{(i+1)}$ allows us to determine whether a specific function $h \in H_n$ is good for $D_k^{(i+1)}$. This is done by considering all the relevant triples (u, v, w) in $D_k^{(i+1)}$, computing for each such (u, v, w) the three quantities (i.e., probabilities) appearing in Eq. (4.2), and deciding accordingly. Trying all possible $h \in H_n$, we find a function (to be denoted $h^{(i+1)}$) that is good for $D_k^{(i+1)}$. This is done while using an additional storage of $s' = O(n + \log|D'_k|)$ (on top of the storage used to record $h^{(1)}, ..., h^{(i)}$), and in time that is exponential in s'. Thus, given D'_k, *we find a good sequence of hash functions,* $h^{(1)}, ..., h^{(t)}$, *in time exponential in* s' *and while using space* $s' + t \cdot \log_2|H_n| = O(t \cdot s')$. Such a sequence of functions allows us to emulate edge-traversals on $D_k^{(t+1)}$, which in turn allows us to (deterministically) approximate the probability that D'_k accepts a random input (i.e., the probability that, starting at the single source vertex of the first layer, automaton D'_k reaches some accepting vertex at the last layer). This approximation is obtained by computing the corresponding probability in $D_k^{(t+1)}$ by traversing all 2^n edges.

To summarize, given D'_k, we can (deterministically) approximate the probability that D'_k accepts a random input in $O(t \cdot s')$-space and $\exp(O(s' + n))$-time, where

$s' = O(n + \log |D'_k|)$ and $t < \log_2 |D'_k|$. Recalling that $n = \Theta(\log |D'_k|)$, this means $O(\log |D'_k|)^2$-space and poly($|D'_k|$)-time. We comment that the approximation can be made accurate up to an additive error term of $1/\text{poly}(|D'_k|)$, whereas the derandomization can tolerate any additive error smaller than $1/6$.

Notes

As stated in the first paper on the subject of "space-resilient pseudorandom generators" [2],[8] this research direction was inspired by the derandomization result obtained via the use of general-purpose pseudorandom generators. The latter result (necessarily) depends on intractability assumptions, and so the objective was identifying natural classes of algorithms for which derandomization is possible without relying on intractability assumptions (but rather by relying on intractability results that are known for the corresponding classes of distinguishers). This objective was achieved before for the case of constant-depth (randomized) circuits [49], but space-bounded (randomized) algorithms offer a more appealing class that refers to natural algorithms. Fundamentally different constructions of space-resilient pseudorandom generators were given in several works, but are superseded by the two incomparable results mentioned in Section 4.2: Theorem 4.2 (a.k.a Nisan's Generator [50]) and Theorem 4.3 (a.k.a the Nisan–Zuckerman Generator [53]). These two results have been "interpolated" in [7]. Theorem 4.4 ($\mathcal{BPL} \subseteq \mathcal{SC}$) was proved by Nisan [51].

We mention that a few years ago, Reingold proved that undirected connectivity can be decided by (deterministic) algorithms of logarithmic space [56]. Prior to his result, only a randomized algorithm of logarithmic space was known (see Appendix D.3).

Exercises

Exercise 4.1 (bounds on the stretch of (s, ε)-pseudorandom generators) Referring to Definition 4.1, establish the following upper-bounds on the stretch ℓ of (s, ε)-pseudorandom generators.

1. If $s(k) \geq 2$ and $\varepsilon(k) \leq 1/2$, then $\ell(k) < \varepsilon(k) \cdot (k+2) \cdot 2^{k+2-s(k)}$.

2. For every $s(k) \geq 1$ and $\varepsilon(k) < 1$ it holds that $\ell(k) < 2^k$.

Guideline: Part 2 follows by combining Exercises 5.11 and 5.12. For Part 1, consider towards the contradiction a generator of stretch $\ell(k) = \varepsilon(k) \cdot (k+2) \cdot 2^{k+2-s(k)}$ and an enumeration, $\alpha^{(1)}, ..., \alpha^{(2^k)} \in \{0,1\}^{\ell(k)}$, of all 2^k outputs of the generator (on k-bit long seeds). Construct a non-uniform automaton of space s that accepts $x_1 \cdots x_{\ell(k)} \in \{0,1\}^{\ell(k)}$ if for some $i \in [\ell(k)/(k+2)]$ it holds that $x_{(i-1)\cdot(k+2)+1} \cdots x_{i\cdot(k+2)}$ equals some string in S_i, where S_i contains the projection of the strings $\alpha^{((i-1)\cdot 2^{s(k)-1}+1)}, ..., \alpha^{(i \cdot 2^{s(k)-1})}$ on the coordinates $(i-1)\cdot(k+2)+1, ..., i\cdot(k+2)$. Note that such an automaton accepts at least $(\ell(k)/(k+2)) \cdot 2^{s(k)-1} = 2\varepsilon(k) \cdot 2^k$ of the possible outputs of the generator, whereas a random ($\ell(k)$-bit long) string is accepted with probability at most $(\ell(k)/(k+2)) \cdot 2^{(s(k)-1)-(k+2)} = \varepsilon(k)/2$.

[8] Interestingly, this paper is more frequently cited for the Expander Random Walk technique, which it has introduced.

Exercise 4.2 (on the existence of (s,ε)-pseudorandom generators) For any s and ε such that $s(k) < k - 2\log_2(k/\varepsilon(k)) - O(1)$, prove the existence of (non-efficient) (s,ε)-pseudorandom generators of stretch $\ell(k) = \Omega(\varepsilon(k)^2 \cdot 2^{k-s(k)}/s(k))$.

Guideline: Use the Probabilistic Method as in Exercise 1.3. Note that non-uniform automata of space s and time ℓ can be described by strings of length $\ell \cdot 2s2^s$.

Exercise 4.3 (multiple samples and space-bounded distinguishers) Let $\{X_k\}_{k\in\mathbb{N}}$ and $\{Y_k\}_{k\in\mathbb{N}}$ be two probability ensembles that are (s,ε)-indistinguishable by non-uniform automata (i.e., the distinguishability-gap of any non-uniform automaton of space s is bounded by the function ε). Then, for any function $t : \mathbb{N}\to\mathbb{N}$, prove that the ensembles $\{(X_k^{(1)}, ..., X_k^{(t(k))})\}_{k\in\mathbb{N}}$ and $\{(Y_k^{(1)}, ..., X_k^{(t(k))})\}_{k\in\mathbb{N}}$ are $(s, t\varepsilon)$-indistinguishable, where $X_k^{(1)}$ through $X_k^{(t(k))}$ and $Y_k^{(1)}$ through $Y_k^{(t(k))}$ are independent random variables, with each $X_k^{(i)}$ identical to X_k and each $Y_k^{(i)}$ identical to Y_k.

Guideline: Use the hybrid technique. When distinguishing the ith and $(i+1)^{\text{st}}$ hybrids, note that the first i blocks (i.e., copies of X_k) as well as the last $t(k) - (i+1)$ blocks (i.e., copies of Y_k) can be fixed and hard-wired into the non-uniform distinguisher.

Exercise 4.4 Provide a more explicit description of the generator outlined in the proof of Theorem 4.2.

Guideline: for $r \in \{0,1\}^n$ and $h^{(1)}, ..., h^{(t)} \in H_n$, the generator outputs a 2^t-long sequence of n-bit strings such that the ith string in this sequence equals $h'(r)$, where h' is a composition of some of the $h^{(j)}$'s.

Chapter 5

Special Purpose Generators

The pseudorandom generators considered so far were aimed at decreasing the amount of randomness utilized by any algorithm of certain time and/or space complexity (or even fully derandomizing the corresponding complexity class). For example, we considered the derandomization of classes such as $\mathcal{BPP}$ and $\mathcal{BPL}$. In the current chapter our goal is less ambitious. We only seek to derandomize (or decrease the randomness of) specific algorithms or rather classes of algorithms that use their random bits in certain (restricted) ways. For example, the algorithm's correctness may only require that its sequence of coin tosses (or "blocks" in such a sequence) are pairwise independent. Indeed, the restrictions that we shall consider here have a concrete and "structural" form, rather than the abstract complexity theoretic forms considered in previous chapters.

The aforementioned restrictions induce corresponding classes of very restricted distinguishers, which in particular are much weaker than the classes of distinguishers considered in previous chapters. These very restricted types of distinguishers induce correspondingly weak types of pseudorandom generators (which produce sequences that fool these distinguishers). Still, such generators have many applications (both in complexity theory and in the design of algorithms).

We start with the simplest of these generators: the pairwise independence generator, and its generalization to t-wise independence for any $t \geq 2$. Such generators *perfectly* fool any distinguisher that only observe t locations in the output sequence. This leads naturally to almost pairwise (or t-wise) independence generators, which also fool such distinguishers (albeit non-perfectly). The latter generators are implied by a stronger class of generators, which is of independent interest: the small-bias generators. Small-bias generators fool any linear test (i.e., any distinguisher that merely considers the XOR of some fixed locations in the input sequence). We finally turn to the Expander Random Walk Generator: This generator produces a sequence of strings that hit any dense subset of strings with probability that is close to the hitting probability of a truly random sequence.[1]

Comment regarding our parameterization: To maintain consistency with prior chapters, we continue to present the generators in terms of the seed length,

[1] Related notions such as samplers, dispersers, and extractors are not treated here (although they were treated in [21, Sec. 3.6] and [24, Apdx. D.3&D.4]).

denoted k. Since this is not the common presentation for most results presented in the sequel, we provide (in footnotes) the common presentation in which the seed length is determined as a function of other parameters.

5.1 Pairwise Independence Generators

Pairwise (resp., t-wise) independence generators fool tests that inspect only two (resp., t) elements in the output sequence of the generator. Such local tests are indeed very restricted, yet they arise naturally in many settings. For example, such a test corresponds to a probabilistic analysis (of a procedure) that only relies on the pairwise independence of certain choices made by the procedure. We also mention that, in some natural range of parameters, pairwise independent sampling is as good as sampling by totally independent sample points (see, e.g., [24, Apdx. D.1.2.4]).

A t-wise independence generator of block-length $b:\mathbb{N}\to\mathbb{N}$ (and stretch function ℓ) is a relatively efficient deterministic algorithm (e.g., one that works in time polynomial in the output length) that expands a k-bit long random seed into a sequence of $\ell(k)/b(k)$ blocks, each of length $b(k)$, such that any t blocks are uniformly and independently distributed in $\{0,1\}^{t\cdot b(k)}$. That is, denoting the ith block of the generator's output (on seed s) by $G(s)_i$, we require that for every $i_1 < i_2 < \cdots < i_t$ (in $[\ell(k)/b(k)]$) it holds that

$$G(U_k)_{i_1}, G(U_k)_{i_2}, ..., G(U_k)_{i_t} \equiv U_{t\cdot b(k)}. \tag{5.1}$$

We note that this condition holds even if the inspected t blocks are selected adaptively (see Exercise 5.1). In case $t=2$, we call the generator pairwise independent.

5.1.1 Constructions

In the first construction, we refer to $\mathrm{GF}(2^{b(k)})$, the finite field of $2^{b(k)}$ elements, and associate its elements with $\{0,1\}^{b(k)}$.

Proposition 5.1 (t-wise independence generator):[2] *Let t be a fixed integer and let $b,\ell,\ell':\mathbb{N}\to\mathbb{N}$ such that $b(k)=k/t$, $\ell'(k)=\ell(k)/b(k)>t$ and $\ell'(k)\le 2^{b(k)}$. Let $\alpha_1,...,\alpha_{\ell'(k)}$ be fixed distinct elements of the field $\mathrm{GF}(2^{b(k)})$. For $s_0,s_1,...,s_{t-1}\in\{0,1\}^{b(k)}$, let*

$$G(s_0,s_1,...,s_{t-1}) \stackrel{\text{def}}{=} \left(\sum_{j=0}^{t-1} s_j\alpha_1^j\,,\ \sum_{j=0}^{t-1} s_j\alpha_2^j\,,...,\ \sum_{j=0}^{t-1} s_j\alpha_{\ell'(k)}^j\right) \tag{5.2}$$

where the arithmetic is that of $\mathrm{GF}(2^{b(k)})$. Then, G is a t-wise independence generator of block-length b and stretch ℓ.

That is, given a seed that consists of t elements of $\mathrm{GF}(2^{b(k)})$, the generator outputs a sequence of $\ell'(k)$ such elements. The proof of Proposition 5.1 is left as an exercise (see Exercise 5.2). It is based on the observation that, for any fixed $v_0,v_1,...,v_{t-1}$,

[2] In the common presentation of this t-wise independence generator, the length of the seed is determined as a function of the desired block-length and stretch. That is, given the parameters b and $\ell'\le 2^b$, the seed length is set to $t\cdot b$.

the condition $\{G(s_0, s_1, ..., s_{t-1})_{i_j} = v_j\}_{j=0}^{t-1}$ constitutes a system of t linear equations over $\mathrm{GF}(2^{b(k)})$ (in the variables $s_0, s_1, ..., s_{t-1}$) such that the equations are linearly-independent. (Thus, linear independence of certain expressions yields statistical independence of the corresponding random variables.)

A somewhat tedious comment. We warn that Eq. (5.2) does not provide a fully explicit construction (of a generator). What is missing is an explicit representation of $\mathrm{GF}(2^{b(k)})$, which requires an irreducible polynomial of degree $b(k)$ over GF(2). For specific values of $b(k)$, a good representation does exist; e.g., for $d \stackrel{\text{def}}{=} b(k) = 2 \cdot 3^e$ (with e being an integer), the polynomial $x^d + x^{d/2} + 1$ is irreducible over GF(2).

We note that a construction analogous to Eq. (5.2) works for every finite field (e.g., a finite field of any prime cardinality), but the problem of providing an explicit representation of such a field remains non-trivial also in other cases (e.g., consider the problem of finding a prime number of size approximately $2^{b(k)}$). The latter fact is the main motivation for considering the following alternative construction for the case of $t = 2$.

The following construction uses (random) affine transformations (as possible seeds). In fact, better performance (i.e., shorter seed length) is obtained by using affine transformations affected by Toeplitz matrices. A Toeplitz matrix is a matrix with all diagonals being homogeneous (see Figure 5.1); that is, $T = (t_{i,j})$ is a Toeplitz matrix if $t_{i,j} = t_{i+1,j+1}$ for all i, j. Note that a Toeplitz matrix is determined by its first row and first column (i.e., the values of $t_{1,j}$'s and $t_{i,1}$'s).

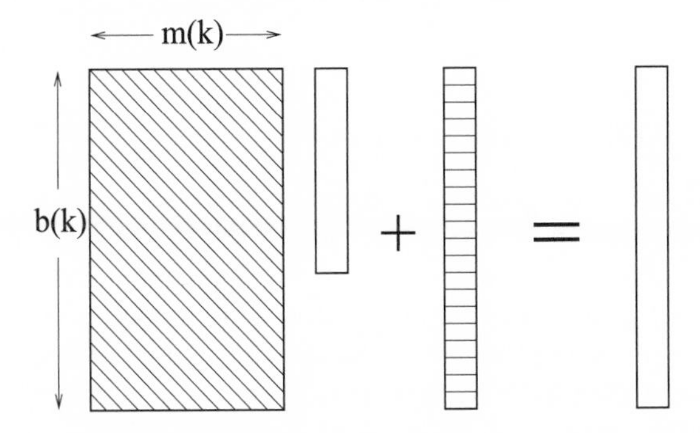

Figure 5.1: An affine transformation affected by a Toeplitz matrix.

Proposition 5.2 (alternative pairwise independence generator, see Figure 5.1):[3] *Let $b, \ell, \ell', m : \mathbb{N} \rightarrow \mathbb{N}$ such that $\ell'(k) = \ell(k)/b(k)$ and $m(k) = \lceil \log_2 \ell'(k) \rceil = k - 2b(k) + 1$. Associate $\{0,1\}^n$ with the n-dimensional vector space over* GF(2)*, and let $v_1, ..., v_{\ell'(k)}$ be fixed distinct vectors in the $m(k)$-dimensional vector space over* GF(2)*. For $s \in \{0,1\}^{b(k)+m(k)-1}$ and $r \in \{0,1\}^{b(k)}$, let*

$$G(s, r) \stackrel{\text{def}}{=} (T_s v_1 + r\,,\, T_s v_2 + r\,, ...,\, T_s v_{\ell'(k)} + r) \tag{5.3}$$

[3] In the common presentation of this pairwise independence generator, the length of the seed is determined as a function of the desired block-length and stretch. That is, given the parameters b and ℓ', the seed length is set to $2b + \lceil \log_2 \ell' \rceil - 1$.

where T_s is a $b(k)$-by-$m(k)$ Toeplitz matrix specified by the string s. Then, G is a pairwise independence generator of block-length b and stretch ℓ.

That is, given a seed that represents an affine transformation defined by a $b(k)$-by-$m(k)$ Toeplitz matrix and a $b(k)$-dimensional vector, the generator outputs a sequence of $\ell'(k) \leq 2^{m(k)}$ strings, each of length $b(k)$. Note that $k = 2b(k)+m(k)-1$, and that the stretching property requires $\ell'(k) > k/b(k)$. The proof of Proposition 5.2 is left as an exercise (see Exercise 5.3). This proof is also based on the observation that linear independence of certain expressions yields statistical independence of the corresponding random variables: here $\{G(s,r)_{i_j} = v_j\}_{j=1}^{2}$ is a system of $2b(k)$ linear equations over GF(2) (in Boolean variables representing the bits of s and r) such that the equations are linearly-independent. We mention that a construction analogous to Eq. (5.3) works for every finite field.

A stronger notion of efficient generation. Ignoring the issue of finding a representation for a large finite field, both the foregoing constructions are efficient in the sense that the generator's output can be produced in time that is polynomial in its length. Actually, the aforementioned constructions satisfy a stronger notion of efficient generation, which is useful in several applications. Specifically, there exists a polynomial-time algorithm that given a seed, $s \in \{0,1\}^k$, and a block location $i \in [\ell'(k)]$ (in binary), outputs the ith block of the corresponding output (i.e., the ith block of $G(s)$). Note that, in the case of the first construction (captured by Eq. (5.2)), this stronger notion depends on the ability to find a representation of $\text{GF}(2^{b(k)})$ in poly(k)-time.[4] Recall that this is possible in the case that $b(k)$ is of the form $2 \cdot 3^e$.

5.1.2 A taste of the applications

Pairwise independence generators do suffice for a variety of applications (cf., [72]). Many of these applications are based on the fact that "Laws of Large Numbers" hold for sequences of trials that are pairwise independent (rather than totally independent). This fact stems from the application of Chebyshev's Inequality, and is the basis of the (rather generic) application to ("pairwise independent") sampling. As a concrete example, we mention the derandomization of a fast parallel algorithm for the Maximal Independent Set problem (as presented in [47, Sec. 12.3]).[5] In general, whenever the analysis of a randomized algorithm only relies on the hypothesis that some objects are distributed in a pairwise independent manner, we may replace its random choices by a sequence of choices that is generated by a pairwise independence generator. Thus, pairwise independence generators suffice for fooling distinguishers that are derived from some natural and interesting randomized algorithms.

Referring to Eq. (5.2), we remark that, for any constant $t \geq 2$, the cost of derandomization (i.e., going over all 2^k possible seeds) is exponential in the block-length (because $b(k) = k/t$). On the other hand, the number of blocks is at most

[4] For the basic notion of efficiency, it suffices to find a representation of $\text{GF}(2^{b(k)})$ in poly($\ell(k)$)-time, which can be done by an exhaustive search in the case that $b(k) = O(\log \ell(k))$.

[5] The core of this algorithm is picking each vertex with probability that is inversely proportional to the vertex's degree. The analysis only requires that these choices be pairwise independent. Furthermore, these choices can be (approximately) implemented by uniformly selecting values in a sufficiently large set.

exponential in the block-length (because $\ell'(k) \leq 2^{b(k)}$), and so if a larger number of blocks is needed, then we can artificially increase the block-length in order to accommodate this (i.e., set $b(k) = \log_2 \ell'(k)$). Thus, the cost of derandomization is polynomial in $\max(\ell'(k), 2^{b'(k)})$, where $\ell'(k)$ denotes the desired number of blocks and $b'(k)$ the desired block-length. (In other words, $\ell'(k)$ denotes the desired number of random choices, and $2^{b'(k)}$ represents the size of the domain of each of these choices.) It follows that *whenever the analysis of a randomized algorithm can be based on a constant amount of independence* between feasibly-many random choices, each taken within a domain of feasible size, *then a feasible derandomization is possible.*

5.2 Small-Bias Generators

As stated in Section 5.1.2, $O(1)$-wise independence generators allow for the efficient derandomization of any efficient randomized algorithm the analysis of which is only based on a *constant amount of independence* between the bits of its random-tape. This restriction is due to the fact that t-wise independence generators of stretch ℓ require a seed of length $\Omega(t \cdot \log \ell)$. Trying to go beyond constant-independence in such derandomizations (while using seeds of length that is logarithmic in the length of the pseudorandom sequence) was the original motivation of the notion of small-bias generators. Specifically, as we shall see in Section 5.2.2, small-bias generators yield meaningful approximations of t-wise independence sequences (based on logarithmic-length seeds).

While the aforementioned type of derandomizations remains an important application of small-bias generators, the latter are of independent interest and have found numerous other applications. In particular, small-bias generators fool "global tests" that examine the entire output sequence and not merely a fixed number of positions in it (as in the case of limited independence generators). Specifically, a small-bias generator produces a sequence of bits that fools any linear test (i.e., a test that computes a fixed linear combination of the bits).

For $\varepsilon : \mathbb{N} \to [0,1]$, an **$\varepsilon$-bias generator** with stretch function ℓ is a relatively efficient deterministic algorithm (e.g., working in $\text{poly}(\ell(k))$-time) that expands a k-bit long random seed into a sequence of $\ell(k)$ bits such that for any fixed non-empty set $S \subseteq \{1, ..., \ell(k)\}$ the bias of the output sequence over S is at most $\varepsilon(k)$. The **bias of a sequence** of n (possibly dependent) Boolean random variables $\zeta_1, ..., \zeta_n \in \{0,1\}$ **over a set** $S \subseteq \{1, ..., n\}$ is defined as

$$2 \cdot \left| \Pr\left[\bigoplus_{i \in S} \zeta_i = 1\right] - \frac{1}{2} \right| = \left| \Pr\left[\bigoplus_{i \in S} \zeta_i = 1\right] - \Pr\left[\bigoplus_{i \in S} \zeta_i = 0\right] \right|. \tag{5.4}$$

The factor of 2 was introduced to make these biases correspond to the Fourier coefficients of the distribution (viewed as a function from $\{0,1\}^n$ to the reals). To see the correspondence replace $\{0,1\}$ by $\{\pm 1\}$, and substitute XOR by multiplication. The bias with respect to a set S is thus written as

$$\left| \Pr\left[\prod_{i \in S} \zeta_i = +1\right] - \Pr\left[\prod_{i \in S} \zeta_i = -1\right] \right| = \left| \mathsf{E}\left[\prod_{i \in S} \zeta_i\right] \right|, \tag{5.5}$$

which is merely the (absolute value of the) Fourier coefficient corresponding to S.

5.2.1 Constructions

Relatively efficient small-bias generators with exponential stretch and exponentially vanishing bias are known.

Theorem 5.3 (small-bias generators):[6] *For some universal constant $c > 0$, let $\ell : \mathbb{N} \to \mathbb{N}$ and $\varepsilon : \mathbb{N} \to [0,1]$ such that $\ell(k) \le \varepsilon(k) \cdot \exp(k/c)$. Then, there exists an ε-bias generator with stretch function ℓ operating in time that is polynomial in the length of its output.*

In particular, we may have $\ell(k) = \exp(k/2c)$ and $\varepsilon(k) = \exp(-k/2c)$. Four simple constructions of small-bias generators that satisfy Theorem 5.3 are known (see [5] and [66, Sec. 3.4]). One of these constructions is based on Linear Feedback Shift Registers (LFSRs), where the seed of the generator is used to determine both the "feedback rule" and the "start sequence" of the LFSR. Specifically, a feedback rule of a t-long LFSR is an irreducible polynomial of degree t over GF(2), denoted $f(x) = x^t + \sum_{j=0}^{t-1} f_j x^j$ where $f_0 = 1$, and the (ℓ-bit long) sequence produced by the corresponding LFSR based on the start sequence $s_0 s_1 \cdots s_{t-1} \in \{0,1\}^t$ is defined as $r_0 r_1 \cdots r_{\ell-1}$, where

$$r_i = \begin{cases} s_i & \text{if } i \in \{0,1,...,t-1\}, \\ \sum_{j=0}^{t-1} f_j \cdot r_{i-t+j} & \text{if } i \in \{t, t+1, ..., \ell-1\} \end{cases} \tag{5.6}$$

(see Figure 5.2). As stated previously, in the corresponding small-bias generator the k-bit long seed is used for selecting an *almost* uniformly distributed feedback rule f (i.e., a random irreducible polynomial of degree $t = k/2$) and a uniformly distributed start sequence s (i.e., a random t-bit string).[7] The corresponding $\ell(k)$-bit long output $r = r_0 r_1 \cdots r_{\ell(k)-1}$ is computed as in Eq. (5.6).

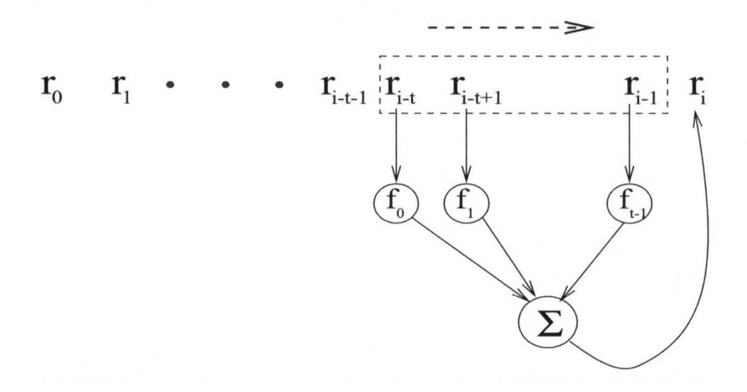

Figure 5.2: The LFSR small-bias generator (for $t = k/2$).

[6] In the common presentation of this generator, the length of the seed is determined as a function of the desired bias and stretch. That is, given the parameters ε and ℓ, the seed length is set to $c \cdot \log(\ell/\varepsilon)$. We comment that using [5] the constant c is merely 2 (i.e., $k \approx 2\log_2(\ell/\varepsilon)$), whereas using [48] $k \approx \log_2 \ell + 4\log_2(1/\varepsilon)$.

[7] Note that an implementation of this generator requires an algorithm for selecting an almost random irreducible polynomial of degree $t = \Omega(k)$. A simple algorithm proceeds by enumerating all irreducible polynomials of degree t, and selecting one of them at random. This algorithm can be implemented (using t random bits) in $\exp(t)$-time, which is $\text{poly}(\ell(k))$ if $\ell(k) = \exp(\Omega(k))$. A $\text{poly}(t)$-time algorithm that uses $O(t)$ random bits is described in [5, Sec. 8].

A stronger notion of efficient generation. As in Section 5.1.1, we note that the aforementioned constructions satisfy a stronger notion of efficient generation, which is useful in several applications. That is, there exists a polynomial-time algorithm that given a k-bit long seed and a bit location $i \in [\ell(k)]$ (in binary), outputs the ith bit of the corresponding output. (For details, see Exercise 5.10.)

5.2.2 A taste of the applications

An archetypical application of small-bias generators is for producing short and random "fingerprints" (or "digests") of strings such that equality and inequality among strings is (probabilistically) reflected in equality and inequality between their corresponding fingerprints. The key observation is that checking whether or not $x = y$ is probabilistically reducible to checking whether the inner product modulo 2 of x and r equals the inner product modulo 2 of y and r, *where r is produced by a small-bias generator G*. Thus, the pair (s, v), where s is a random seed to G and v equals the inner product modulo 2 of z and $G(s)$, serves as the randomized fingerprint of the string z. One advantage of this reduction is that only a few bits (i.e., the seed of the generator and the result of the inner product) need to be "communicated between x and y" in order to enable the checking (see Exercise 5.6). A related advantage is the low randomness complexity of this reduction, which uses $|s|$ rather than $|G(s)|$ random bits, where $|s|$ may be $O(\log|G(s)|)$. This low (i.e., logarithmic) randomness-complexity underlies the application of small-bias generators to the construction of PCP systems and amplifying reductions of gap problems regarding the satisfiability of systems of equations (see, e.g., [24, Exer. 10.6]).

Small-bias generators have been used in a variety of areas (e.g., inapproximation, structural complexity, and applied cryptography; see the references in [21, Sec. 3.6.2]). In addition, as shown next, small-bias generators seem an important tool in the design of various types of "pseudorandom" objects.

Approximate independence generators. As hinted at the beginning of this section, small-bias is related to approximate versions of limited independence.[8] Actually, as implied by Exercise 5.7, even a restricted type of ε-bias (in which only subsets of size $t(k)$ are required to have bias upper-bounded by ε) implies that any $t(k)$ bits in the said sequence are $2^{t(k)/2} \cdot \varepsilon(k)$-close to $U_{t(k)}$, where here we refer to the variation distance (i.e., L1-Norm distance) between the two distributions. (The max-norm of the difference is bounded by $\varepsilon(k)$.)[9] Combining Theorem 5.3 and the foregoing upper-bound, we obtain *generators with exponential stretch* (i.e., $\ell(k) = \exp(\Omega(k))$) that produce *sequences that are approximately $\Omega(k)$-wise independent in the sense that any $t(k) = \Omega(k)$ bits in them are $2^{-\Omega(k)}$-close to $U_{t(k)}$*. Thus, whenever the analysis of a randomized algorithm can be based on a logarithmic amount of (almost) independence between feasibly-many binary random choices, a feasible derandomization is possible (by using an adequate generator of logarithmic seed length).[10]

---

[8] We warn that, unlike in the case of perfect independence, here we refer only to the distribution on fixed bit locations. See Exercise 5.5 for further discussion.

[9] Both bounds are derived from the L2-Norm bound on the difference vector (i.e., the difference between the two probability vectors). For details, see Exercise 5.7.

[10] Furthermore, as shown in Exercise 5.14, relying on the linearity of the construction presented in Proposition 5.1, we can obtain *generators with double-exponential stretch* (i.e., $\ell(k) = \exp(2^{\Omega(k)})$)

Extensions to non-binary choices were considered in various works (see references in [21, Sec. 3.6.2]). Some of these works also consider the related problem of constructing small "discrepancy sets" for geometric and combinatorial rectangles.

t-universal set generators. Using the aforementioned upper-bound on the max-norm (of the deviation from uniform of any t locations), any ε-bias generator yields a *t-universal set generator*, provided that $\varepsilon < 2^{-t}$. The latter generator outputs sequences such that in every subsequence of length t all possible 2^t patterns occur (i.e., each for at least one possible seed). Such generators have many applications.

5.2.3 Generalization

In this section, we outline a generalization of the treatment of small-bias generators to the generation of sequences over an arbitrary finite field. Focusing on the case of a field of prime cardinality, denoted GF(p), we first define an adequate notion of bias. Generalizing Eq. (5.5), we define the bias of a sequence of n (possibly dependent) random variables $\zeta_1, ..., \zeta_n \in \mathrm{GF}(p)$ with respect to the linear combination $(c_1, ..., c_n) \in \mathrm{GF}(p)^n$ as $\left\|\mathsf{E}\left[\omega^{\sum_{i=1}^n c_i \zeta_i}\right]\right\|$, where ω denotes the pth (complex) root of unity (i.e., $\omega = -1$ if $p = 2$). Referring to Exercise 5.16, we note that upper-bounds on the biases of $\zeta_1, ..., \zeta_n$ (with respect to any non-zero linear combinations) yield upper-bounds on the distance of $\sum_{i=1}^n c_i \zeta_i$ from the uniform distribution over GF(p).

We say that $S \subseteq \mathrm{GF}(p)^n$ is an ε-bias probability space if a uniformly selected sequence in S has bias at most ε with respect to any non-zero linear combination over GF(p). (Whenever such a space is efficiently constructible, it yields a corresponding ε-biased generator.) We mention that the LFSR construction, outlined in Section 5.2.1 and analyzed in Exercise 5.9, generalizes to GF(p) and yields an ε-bias probability space of size (at most) p^{2e}, where $e = \lceil \log_p(n/\varepsilon) \rceil$. Such constructions can be used in applications that generalize those in Section 5.2.2.

A different generalization. Recalling that small-bias generators fool all linear tests, we consider generators that fool any test that can be represented by a polynomial of degree d. It was recently proved that taking the sum of d independently distributed outputs produced by a small-bias generator (on d independently chosen seeds) yields a sequence that fools all degree d tests [70]. (Interestingly, this sequence may not fool all polynomials of degree $d+1$; see [66].)

5.3 Random Walks on Expanders

In this section we review generators that produce a sequence of values by taking a random walk on a large graph that has a small degree but an adequate "mixing" property (in the sense that a random walk of logarithmic length that starts at any fixed vertex reaches an almost uniformly distributed vertex). Such a graph is called an expander, and by taking a random walk (of length ℓ') on it we generate a sequence

that are approximately $t(k)$-independent (in the foregoing sense). That is, we may obtain generators with stretch $\ell(k) = 2^{2^{\Omega(k)}}$ producing bit sequences in which any $t(k) = \Omega(k)$ positions have variation distance at most $\varepsilon(k) = 2^{-\Omega(k)}$ from uniform; in other words, such generators may have seed-length $k = O(t(k) + \log(1/\varepsilon(k)) + \log\log \ell(k))$. In the corresponding result for the max-norm distance, it suffices to have $k = O(\log(t(k)/\varepsilon(k)) + \log\log \ell(k))$.

of ℓ' values over its vertex set, while using a random seed of length $b+(\ell'-1)\cdot\log_2 d$, where 2^b denotes the number of vertices in the graph and d denotes its degree. This seed length should be compared against the $\ell'\cdot b$ random bits required for generating a sequence of ℓ' independent samples from $\{0,1\}^b$ (or taking a random walk on a clique of size 2^b). Interestingly, as we shall see, the pseudorandom sequence (generated by the said random walk on an expander) *behaves similarly to a truly random sequence with respect to hitting any dense subset of* $\{0,1\}^b$. Let us start by defining this property (or rather by defining the corresponding hitting problem).

Definition 5.4 (the hitting problem): *A sequence of* (possibly dependent) *random variables, denoted* $(X_1,...,X_{\ell'})$, *over* $\{0,1\}^b$ *is* (ε,δ)-hitting *if for any* (target) *set* $T\subseteq\{0,1\}^b$ *of cardinality at least* $\varepsilon\cdot 2^b$, *with probability at least* $1-\delta$, *at least one of these variables hits* T; *that is,* $\Pr[\exists i \text{ s.t. } X_i\in T]\geq 1-\delta$.

Clearly, a truly random sequence of length ℓ' over $\{0,1\}^b$ is (ε,δ)-hitting for $\delta=(1-\varepsilon)^{\ell'}$. The aforementioned "expander random walk generator" (to be described next) achieves similar behavior.[11] Specifically, for arbitrary small $c>0$ (which depends on the degree and the mixing property of the expander), the generator's output is (ε,δ)-hitting for $\delta=(1-(1-c)\cdot\varepsilon)^{\ell'}$. To describe this generator, we need to discuss expanders.

5.3.1 Background: expanders and random walks on them

By expander graphs (or expanders) of degree d and eigenvalue bound $\lambda<d$, we actually mean an infinite family of d-regular[12], graphs, $\{G_N\}_{N\in\mathbb{S}}$ ($\mathbb{S}\subseteq\mathbb{N}$), such that G_N is a d-regular graph over N vertices and the absolute value of all eigenvalues, save the biggest one, of the adjacency matrix of G_N is upper-bounded by λ. For simplicity, we shall assume that the vertex set of G_N is $[N]$ (although in some constructions a somewhat more redundant representation is more convenient). We will refer to such a family as a (d,λ)-expander (for $\mathbb{S}$). This technical definition is related to the aforementioned notion of "mixing" (which refers to the rate at which a random walk starting at a fixed vertex reaches uniform distribution over the graph's vertices).

We are interested in explicit constructions of such graphs, by which we mean that there exists a polynomial-time algorithm that on input N (in binary), a vertex v in G_N and an index $i\in\{1,...,d\}$, returns the ith neighbor of v. (We also require that the set $\mathbb{S}$ for which G_N's exist is sufficiently "tractable" – say, that given any $n\in\mathbb{N}$ one may efficiently find an $s\in\mathbb{S}$ such that $n\leq s<2n$.) Several explicit constructions of expanders are known (cf., e.g., [44, 43, 57]). Below, we rely on the fact that for every $\overline{\lambda}>0$, there exist d and an explicit construction of a $(d,\overline{\lambda}\cdot d)$-expander over $\{2^b : b\in\mathbb{N}\}$.[13] The relevant (to us) fact about expanders is stated next.

Theorem 5.5 (Expander Random Walk Theorem): *Let* $G=(V,E)$ *be an expander graph of degree* d *and eigenvalue bound* λ. *Consider taking a random walk on* G *by uniformly selecting a start vertex and taking* $\ell'-1$ *additional random steps such that*

---

[11] We comment that other pseudorandom generators that were considered in this text also exhibit hitting properties; see Exercise 5.17.

[12] A graph is called d-regular if each of its vertices has exactly d neighbors.

[13] This can be obtained with $d=\text{poly}(1/\overline{\lambda})$. In fact, $d=O(1/\overline{\lambda}^2)$, which is optimal, can be obtained too, albeit with graphs of sizes that are only approximately powers of two.

at each step the walk uniformly selects an edge incident at the current vertex and traverses it. Then, for any $W \subseteq V$ and $\rho \stackrel{\rm def}{=} |W|/|V|$, the probability that such a random walk stays in W is at most

$$\rho \cdot \left(\rho + (1-\rho)\cdot\frac{\lambda}{d}\right)^{\ell'-1} . \tag{5.7}$$

Thus, a random walk on an expander is "pseudorandom" with respect to the hitting property (i.e., when we consider hitting the set $V \setminus W$ and use $\varepsilon = 1-\rho$); that is, a set of density ε is hit with probability at least $1-\delta$, where $\delta = (1-\varepsilon)\cdot(1-\varepsilon+(\lambda/d)\cdot\varepsilon)^{\ell'-1} < (1-(1-(\lambda/d))\cdot\varepsilon)^{\ell'}$. A proof of Theorem 5.5 is given in [36], while a proof of an upper-bound that is weaker than Eq. (5.7) is outlined next.

A weak version of the Expander Random Walk Theorem: Using notation as in Theorem 5.5, we claim that the probability that a random walk of length ℓ' stays in W is at most $(\rho+(\lambda/d)^2)^{\ell'/2}$. In fact, we make a more general claim that refers to the probability that a random walk of length ℓ' intersects $W_0 \times W_1 \times \cdots \times W_{\ell'-1}$. The claimed upper-bound is

$$\sqrt{\rho_0}\cdot\prod_{i=1}^{\ell'-1}\sqrt{\rho_i+(\lambda/d)^2}, \tag{5.8}$$

where $\rho_i \stackrel{\rm def}{=} |W_i|/|V|$. In order to prove Eq. (5.8), we view the random walk as the evolution of a corresponding probability vector under suitable transformations. The transformations correspond to taking a random step in the graph and to passing through a "sieve" that keeps only the entries that correspond to the current set W_i. The key observation is that the first transformation shrinks the component that is orthogonal to the uniform distribution, whereas the second transformation shrinks the component that is in the direction of the uniform distribution. (See Exercise 5.18.)

5.3.2 The generator

Using Theorem 5.5 and an explicit $(2^t, \overline{\lambda}\cdot 2^t)$-expander, we obtain a generator that produces sequences that are (ε,δ)-hitting for δ that is almost optimal.

Proposition 5.6 (The Expander Random Walk Generator):[14] *For every constant $\overline{\lambda} > 0$, consider an explicit construction of $(2^t, \overline{\lambda}\cdot 2^t)$-expanders for $\{2^n : n \in \mathbb{N}\}$, where $t \in \mathbb{N}$ is a sufficiently large constant. For $v \in [2^n] \equiv \{0,1\}^n$ and $i \in [2^t] \equiv \{0,1\}^t$, denote by $\Gamma_i(v)$ the vertex of the corresponding 2^n-vertex graph that is reached from vertex v when following its $i^{\rm th}$ edge. For $b, \ell' : \mathbb{N} \to \mathbb{N}$ such that $k = b(k) + (\ell'(k)-1)\cdot t < \ell'(k)\cdot b(k)$, and for $v_0 \in \{0,1\}^{b(k)}$ and $i_1, ..., i_{\ell'(k)-1} \in [2^t]$, let*

$$G(v_0, i_1, ..., i_{\ell'(k)-1}) \stackrel{\rm def}{=} (v_0, v_1, ..., v_{\ell'(k)-1}), \tag{5.9}$$

where $v_j = \Gamma_{i_j}(v_{j-1})$. Then, G has stretch $\ell(k) = \ell'(k)\cdot b(k)$, and $G(U_k)$ is (ε,δ)-hitting for any $\varepsilon > 0$ and $\delta = (1-(1-\overline{\lambda})\cdot\varepsilon)^{\ell'(k)}$.

[14] In the common presentation of this generator, the length of the seed is determined as a function of the desired block-length and stretch. That is, given the parameters b and ℓ', the seed length is set to $b + (\ell'-1)\cdot t$.

The stretch of G is maximized at $b(k) \approx k/2$ (and $\ell'(k) = k/2t$), but maximizing the stretch is not necessarily the goal in all applications. In many applications, the parameters n, ε and δ are given, and the goal is to derive a generator that produces (ε, δ)-hitting sequences over $\{0,1\}^n$ while minimizing both the length of the sequence and the amount of randomness used by the generator (i.e., the seed length). Indeed, Proposition 5.6 suggests using sequences of length $\ell' \approx \varepsilon^{-1} \log_2(1/\delta)$ that are generated based on a random seed of length $n + O(\ell')$.

Expander random-walk generators have been used in a variety of areas (e.g., PCP and inapproximability (see [10, Sec. 11.1]), cryptography (see [22, Sec. 2.6]), and the design of various types of "pseudorandom" objects.

Notes

The various generators presented in Chapter 5 were not inspired by any of the other types of pseudorandom generator (nor even by the generic notion of pseudorandomness). Pairwise independence generators were explicitly suggested in [15] (and are implicit in [13]). The generalization to t-wise independence (for $t \geq 2$) is due to [4]. Small-bias generators were first defined and constructed by Naor and Naor [48], and three simple constructions were subsequently given in [5]. The Expander Random Walk Generator was suggested by Ajtai, Komlos, and Szemerédi [2], who discovered that random walks on expander graphs provide a good approximation to repeated independent attempts to hit any fixed subset of sufficient density (within the vertex set). The analysis of the hitting property of such walks was subsequently improved, culminating in the bound cited in Theorem 5.5, which is taken from [36, Cor. 6.1].

Exercises

Exercise 5.1 (adaptive t-wise independence tests) Recall that a generator $G : \{0,1\}^k \to \{0,1\}^{\ell'(k) \cdot b(k)}$ is called t-wise independent if *for any t fixed block positions*, the distribution $G(U_k)$ restricted to these t blocks is uniform over $\{0,1\}^{t \cdot b(k)}$. Prove that the output of a t-wise independence generator is (perfectly) indistinguishable from the uniform distribution *by any test that examines t of the blocks, even if the examined blocks are selected adaptively* (i.e., the location of the ith block to be examined is determined based on the contents of the previously inspected blocks).

Guideline: First show that, without loss of generality, it suffices to consider deterministic (adaptive) testers. Next, show that the probability that such a tester sees any fixed sequence of t values at the locations selected *adaptively* (in the generator's output) equals $2^{-t \cdot b(k)}$, where $b(k)$ is the block-length.

Exercise 5.2 (a t-wise independence generator) Prove that G as defined in Proposition 5.1 produces a t-wise independent sequence over $\mathrm{GF}(2^{b(k)})$.

Guideline: For every t fixed sequence of indices $i_1, ..., i_t \in [\ell'(k)]$, consider the distribution of $G(U_k)_{i_1,...,i_t}$ (i.e., the projection of $G(U_k)$ on locations $i_1, ..., i_t$). Show that for every sequence of t possible values $v_1, ..., v_t \in \mathrm{GF}(2^{b(k)})$, there exists a unique seed $s \in \{0,1\}^k$ such that $G(s)_{i_1,...,i_t} = (v_1, ..., v_t)$.

Exercise 5.3 (pairwise independence generators) As a warm-up, consider a construction analogous to the one in Proposition 5.2, except that here the seed specifies an arbitrary affine $b(k)$-by-$m(k)$ transformation. That is, for $s \in \{0,1\}^{b(k)\cdot m(k)}$ and $r \in \{0,1\}^{b(k)}$, where $k = b(k) \cdot m(k) + b(k)$, let

$$G(s,r) \stackrel{\text{def}}{=} (A_s v_1 + r\,,\; A_s v_2 + r\,, ...,\; A_s v_{\ell'(k)} + r) \tag{5.10}$$

where A_s is a $b(k)$-by-$m(k)$ matrix specified by the string s. Show that G as in Eq. (5.10) is a pairwise independence generator of block-length b and stretch ℓ. Next, show that G as in Eq. (5.3) is a pairwise independence generator of block-length b and stretch ℓ.

Guideline: The following description applies to both constructions. First note that for every fixed $i \in [\ell'(k)]$, the ith element in the sequence $G(U_k)$, denoted $G(U_k)_i$, is uniformly distributed in $\{0,1\}^{b(k)}$. Actually, show that for every fixed $s \in \{0,1\}^{k-b(k)}$, it holds that $G(s, U_{b(k)})_i$ is uniformly distributed in $\{0,1\}^{b(k)}$. Next note that it suffices to show that, for every $j \neq i$, conditioned on the value of $G(U_k)_i$, the value of $G(U_k)_j$ is uniformly distributed in $\{0,1\}^{b(k)}$. The key technical detail is showing that, for any non-zero vector $v \in \{0,1\}^{m(k)}$ and a uniformly selected $s \in \{0,1\}^{k-b(k)}$, it holds that $A_s v$ (resp., $T_s v$) is uniformly distributed in $\{0,1\}^{b(k)}$. This is easy in case of a random $b(k)$-by-$m(k)$ matrix, and can be proven also for a random Toeplitz matrix.

Exercise 5.4 In continuation of the warm-up of Exercise 5.3, consider the following construction (which appears in the proof of Theorem 2.11; see Appendix C). For $t > 1$, let $b(k) = k/t$, and consider the mapping of $(s^1, ..., s^t) \in \{0,1\}^{t\cdot b(k)}$ to $(r^J) \in \{0,1\}^{(2^t-1)\cdot b(k)}$, where the J's range over all *non-empty* subsets of $\{1, 2, ..., t\}$ and $r^J \stackrel{\text{def}}{=} \bigoplus_{j\in J} s^j$. Prove that G is a pairwise independence generator of block-length b and stretch $\ell(k) = \frac{2^t-1}{t} \cdot k$.

Guideline: For $J \neq J'$, it holds that $r^J \oplus r^{J'} = \bigoplus_{j\in K} s^j$, where K denotes the symmetric difference of J and J'.

Exercise 5.5 (adaptive t-wise independence tests, revisited) Prove that, in contrast to Exercise 5.1, with respect to *non-perfect* indistinguishability, there is a discrepancy between adaptive and non-adaptive tests that inspect t locations.

1. Specifically, present a distribution over 2^{t-1}-bit long strings in which every t fixed bit positions are $t\cdot 2^{-t}$-close to uniform, but there exists a test that adaptively inspects t positions and distinguishes this distribution from the uniform one with gap of $1/2$.

   Guideline: Modify the uniform distribution over $((t-1) + 2^{t-1})$-bit long strings such that the first $t-1$ locations indicate a bit position (among the rest) that is set to zero.

2. On the other hand, prove that if every t fixed bit positions in a distribution X are ε-close to uniform, then every test that adaptively inspects t positions can distinguish X from the uniform distribution with gap at most $2^t \cdot \varepsilon$.

   Guideline: See Exercise 5.1.

Exercise 5.6 Suppose that G is an ε-bias generator with stretch ℓ. Show that equality between the $\ell(k)$-bit strings x and y can be probabilistically checked (with error probability $(1+\varepsilon)/2$) by comparing the inner product modulo 2 of x and $G(s)$ to the inner product modulo 2 of y and $G(s)$, where $s \in \{0,1\}^k$ is selected uniformly. Note that this method is a randomness-efficient approximation of comparing the inner product modulo 2 of x and r to the inner product modulo 2 of y and r, where $r \in \{0,1\}^{\ell(k)}$ is selected uniformly.
(Hint: Consider the special case in which $y = 0^{\ell(k)}$.)

Exercise 5.7 (bias vs. statistical difference from uniform) Let X be a random variable assuming values in $\{0,1\}^t$. Prove that if X has bias at most ε over any non-empty set then the statistical difference between X and U_t is at most $2^{t/2} \cdot \varepsilon$, and that for every $x \in \{0,1\}^t$ it holds that $\mathsf{Pr}[X = x] = 2^{-t} \pm \varepsilon$.

Guideline: Consider the probability function $p : \{0,1\}^t \to [0,1]$ defined by $p(x) \stackrel{\text{def}}{=} \mathsf{Pr}[X = x]$, and let $\delta(x) \stackrel{\text{def}}{=} p(x) - 2^{-t}$ denote the deviation of p from the uniform probability function. Viewing the set of real functions over $\{0,1\}^t$ as a 2^t-dimensional vector space, consider two orthonormal bases for this space. The first basis consists of the (Kroniker) functions $\{k_\alpha\}_{\alpha \in \{0,1\}^t}$ such that $k_\alpha(x) = 1$ if $x = \alpha$ and $k_\alpha(x) = 0$ otherwise. The second basis consists of the (normalized Fourier) functions $\{f_S\}_{S \subseteq [t]}$ defined by $f_S(x_1 \cdots x_t) \stackrel{\text{def}}{=} 2^{-t/2} \prod_{i \in S} (-1)^{x_i}$ (where $f_\emptyset \equiv 2^{-t/2}$).[15] Note that the bias of X over any $S \neq \emptyset$ equals $|\sum_x p(x) \cdot 2^{t/2} f_S(x)|$, which in turn equals $2^{t/2} |\sum_x \delta(x) f_S(x)|$. Thus, for every S (including the empty set), we have $|\sum_x \delta(x) f_S(x)| \leq 2^{-t/2}\varepsilon$, which means that the representation of δ in the normalized Fourier basis is by coefficients that have each an absolute value of at most $2^{-t/2}\varepsilon$. It follows that the L2-Norm of this vector of coefficients is upper-bounded by $\sqrt{2^t \cdot (2^{-t/2}\varepsilon)^2} = \varepsilon$, and the two claims follow by noting that they refer to norms of δ according to the Kroniker basis. In particular, the L2-Norm is preserved under orthonormal bases, the max-norm is upper-bounded by the L2-Norm, and the L1-Norm is upper-bounded by $\sqrt{2^t}$ times the value of the L2-Norm.

Exercise 5.8 (on the existence of (non-explicit) small-bias generators) Prove that, for $k = \log_2(\ell(k)/\varepsilon(k)^2) + O(1)$, there exists a function $G : \{0,1\}^k \to \{0,1\}^{\ell(k)}$ such that $G(U_k)$ has bias at most $\varepsilon(k)$ over any non-empty subset of $[\ell(k)]$.

Guideline: Use the Probabilistic Method as in Exercise 1.3.

Exercise 5.9 (The LFSR small-bias generator (following [5])) Using the following guidelines (and letting $t = k/2$), analyze the construction outlined following Theorem 5.3 (and depicted in Figure 5.2):

1. Prove that r_i equals $\sum_{j=0}^{t-1} c_j^{(f,i)} \cdot s_j$, where $c_j^{(f,i)}$ is the coefficient of z^j in the (degree $t-1$) polynomial obtained by reducing z^i modulo the polynomial $f(z)$ (i.e., $z^i \equiv \sum_{j=0}^{t-1} c_j^{(f,i)} z^j \pmod{f(z)}$).

   Guideline: Recall that $z^t \equiv \sum_{j=0}^{t-1} f_j z^j \pmod{f(z)}$, and thus for every $i \geq t$ it holds that $z^i \equiv \sum_{j=0}^{t-1} f_j z^{i-t+j} \pmod{f(z)}$. Note the correspondence to $r_i = \sum_{j=0}^{t-1} f_j \cdot r_{i-t+j}$.

[15] Verify that both bases are indeed orthogonal (i.e., $\sum_x k_\alpha(x) k_\beta(x) = 0$ for every $\alpha \neq \beta$ and $\sum_x f_S(x) f_T(x) = 0$ for every $S \neq T$) and normal (i.e., $\sum_x k_\alpha(x)^2 = 1$ and $\sum_x f_S(x)^2 = 1$).

2. For any non-empty $S \subseteq \{0,...,\ell(k)-1\}$, evaluate the bias of the sequence $r_0,...,r_{\ell(k)-1}$ over S, where f is a random irreducible polynomial of degree t and $s = (s_0,...,s_{t-1}) \in \{0,1\}^t$ is uniformly distributed. Specifically:

   (a) For a fixed f and random $s \in \{0,1\}^t$, prove that $\sum_{i\in S} r_i$ has non-zero bias if and only if $f(z)$ divides $\sum_{i\in S} z^i$.

   (Hint: Note that $\sum_{i\in S} r_i = \sum_{j=0}^{t-1} \sum_{i\in S} c_j^{(f,i)} s_j$, and use Item 1.)

   (b) Prove that the probability that a random irreducible polynomial of degree t divides $\sum_{i\in S} z^i$ is $\Theta(\ell(k)/2^t)$.

   (Hint: A polynomial of degree n can be divided by at most n/d different irreducible polynomials of degree d. On the other hand, the number of irreducible polynomials of degree d over GF(2) is $\Theta(2^d/d)$.)

   Conclude that for random f and s, the sequence $r_0,...,r_{\ell(k)-1}$ has bias $O(\ell(k)/2^t)$.

Note that an implementation of the LFSR generator requires a mapping of random $k/2$-bit long string to *almost* random irreducible polynomials of degree $k/2$. Such a mapping can be constructed in $\exp(k)$-time, which is $\text{poly}(\ell(k))$ if $\ell(k) = \exp(\Omega(k))$. A more efficient mapping that uses a $O(k)$-bit long seed is described in [5, Sec. 8].

Exercise 5.10 Show that the LFSR small-bias generator, depicted in Figure 5.2 satisfies a stronger notion of efficient generation; specifically, there exists a polynomial-time algorithm that given a k-bit long seed and a bit location $i \in [\ell(k)]$ (in binary), outputs the ith bit of the corresponding output.

Guideline: The assertion is based on the fact that when this generator is fed with seed $(f_0,...,f_{(k/2)-1}, s_0,...,s_{(k/2)-1})$, its output sequence $(r_0, r_1, ..., r_{\ell(k)})$ satisfies

$$\begin{pmatrix} r_{i-t+1} \\ r_{i-t+2} \\ \vdots \\ r_{i-1} \\ r_i \end{pmatrix} = \begin{pmatrix} 0 & 1 & 0 & \cdots & 0 \\ 0 & 0 & 1 & \cdots & 0 \\ \vdots & \vdots & \vdots & \cdots & \vdots \\ 0 & 0 & 0 & \cdots & 1 \\ f_0 & f_1 & f_2 & \cdots & f_{t-1} \end{pmatrix} \begin{pmatrix} r_{i-t} \\ r_{i-t+1} \\ \vdots \\ r_{i-2} \\ r_{i-1} \end{pmatrix} = \begin{pmatrix} 0 & 1 & 0 & \cdots & 0 \\ 0 & 0 & 1 & \cdots & 0 \\ \vdots & \vdots & \vdots & \cdots & \vdots \\ 0 & 0 & 0 & \cdots & 1 \\ f_0 & f_1 & f_2 & \cdots & f_{t-1} \end{pmatrix}^{i-t+1} \begin{pmatrix} s_0 \\ s_1 \\ \vdots \\ s_{t-2} \\ s_{t-1} \end{pmatrix}.$$

Exercise 5.11 (limitations on small-bias generators) Let G be an ε-bias generator with stretch ℓ, and view G as a mapping from $\text{GF}(2)^k$ to $\text{GF}(2)^{\ell(k)}$. As such, each bit in the output of G can be viewed as a polynomial[16] in the k input variables (each ranging in GF(2)). Prove that if $\varepsilon(k) < 1$ and each of these polynomials has *total degree* at most d, then $\ell(k) \leq \sum_{i=1}^{d} \binom{k}{i}$. Derive the following corollaries:

1. If $\varepsilon(k) < 1$, then $\ell(k) < 2^k$ (regardless of d).[17]

[16] Recall that every Boolean function over GF(p) can be expressed as a polynomial of *individual degree* at most $p-1$.

[17] This upper-bound is optimal, because (efficient) ε-bias generators of stretch $\ell(k) = \text{poly}(\varepsilon(k)) \cdot 2^k$ do exist (see [48]).

2. If $\varepsilon(k) < 1$ and $\ell(k) > k$, then G cannot be a linear transformation.[18]

Guideline (for the main claim): Note that, without loss of generality, all the aforementioned polynomials have a free term equal to zero (and have individual degree at most 1 in each variable). Next, consider the vector space spanned by all d-monomials over k variables (i.e., monomials having at most d variables). Since $\varepsilon(k) < 1$, the polynomials representing the output bits of G must correspond to a sequence of independent vectors in this space.

Exercise 5.12 (a sanity check for space-bounded pseudorandomness) The following fact is suggested as a sanity check for candidate pseudorandom generators with respect to space-bounded automata. The fact (to be proven as an exercise) is that, for every $\varepsilon(\cdot)$ and $s(\cdot)$ such that $s(k) \geq 1$ for every k, if G is (s, ε)-pseudorandom (as per Definition 4.1), then G is an ε-bias generator.

Exercise 5.13 In contrast to Exercise 5.12, prove that there exist $\exp(-\Omega(n))$-bias distributions over $\{0,1\}^n$ that are not $(2, 0.666)$-pseudorandom.

Guideline: Show that the uniform distribution over the set

$$\left\{\sigma_1 \cdots \sigma_n : \sum_{i=1}^{n} \sigma_i \equiv 0 \pmod 3\right\}$$

has bias $\exp(-\Omega(n))$. An alternative construction appears in [66, Sec. 3.5].

Exercise 5.14 (approximate t-wise independence generators (cf. [48])) Combining a small-bias generator as in Theorem 5.3 with the t-wise independence generator of Eq. (5.2), and relying on the linearity of the latter, construct a generator producing ℓ-bit long sequences in which any t positions are at most ε-away from uniform (in variation distance), while using a seed of length $O(t+\log(1/\varepsilon)+\log\log \ell)$. (For max-norm a seed of length $O(\log(t/\varepsilon) + \log\log \ell)$ suffices.)

Guideline: First note that, for any t, ℓ' and $b \geq \log_2 \ell'$, the transformation of Eq. (5.2) can be implemented by a fixed linear (over GF(2)) transformation of a $t \cdot b$-bit seed into an ℓ-bit long sequence, where $\ell = \ell' \cdot b$. It follows that, for $b = \log_2 \ell'$, there exists a fixed GF(2)-linear transformation T of a random seed of length $t \cdot b$ into a t-wise independent bit sequence of the length ℓ (i.e., $T\,U_{t \cdot b}$ is t-wise independent over $\{0,1\}^\ell$). Thus, every t rows of T are linearly independent. The key observation is that when we replace the aforementioned random seed by an ε'-bias sequence, every set of $i \leq t$ positions in the output sequence has bias at most ε' (because they define a non-zero linear test on the bits of the ε'-bias sequence). Note that the length of the new seed (used to produce ε'-bias sequence of length $t \cdot b$) is $O(\log tb/\varepsilon')$. Applying Exercise 5.7, we conclude that any t positions are at most $2^{t/2} \cdot \varepsilon'$-away from uniform (in variation distance). Recall that this was obtained using a seed of length $O(\log(t/\varepsilon') + \log\log \ell)$, and the claim follows by using $\varepsilon' = 2^{-t/2} \cdot \varepsilon$.

Exercise 5.15 (small-bias generator and error-correcting codes) Show a correspondence between ε-bias generators of stretch ℓ and binary linear error-correcting

[18] In contrast, bilinear ε-bias generators (i.e., with $\ell(k) > k$) do exist; for example, $G(s) = (s, b(s))$, where $b(s_1, ..., s_k) = \sum_{i=1}^{k/2} s_i s_{(k/2)+i} \bmod 2$, is an ε-bias generator with $\varepsilon(k) = \exp(-\Omega(k))$. (Hint: Focusing on bias over sets that include the last output bit, prove that, without loss of generality, it suffices to analyze the bias of $b(U_k)$.)

codes mapping $\ell(k)$-bit long strings to 2^k-bit long strings such that every two codewords are at distance $(1 \pm \varepsilon(k)) \cdot 2^{k-1}$ apart.

Guideline: Associate $\{0,1\}^k$ with $[2^k]$. Then, a generator $G : [2^k] \to \{0,1\}^{\ell(k)}$ corresponds to the code $C : \{0,1\}^{\ell(k)} \to \{0,1\}^{2^k}$ such that, for every $i \in [\ell(k)]$ and $j \in [2^k]$, the ith bit of $G(j)$ equals the jth bit of $C(0^{i-1}10^{\ell(k)-i})$.

Exercise 5.16 (on the bias of sequences over a finite field) For a prime p, let ζ be a random variable assigned values in $\mathrm{GF}(p)$ and $\delta(v) \stackrel{\text{def}}{=} \Pr[\zeta = v] - (1/p)$. Prove that $\max_{v \in \mathrm{GF}(p)}\{|\delta(v)|\}$ is upper-bounded by $b \stackrel{\text{def}}{=} \max_{c \in \{1,\dots,p-1\}}\{\|\mathsf{E}[\omega^{c\zeta}]\|\}$, where ω denotes the pth (complex) root of unity, and that $\sum_{v \in \mathrm{GF}(p)} |\delta(v)|$ is upper-bounded by $\sqrt{p} \cdot b$.

Guideline: Analogously to Exercise 5.7, view probability distributions over $\mathrm{GF}(p)$ as p-dimensional vectors, and consider two bases for the set of complex functions over $\mathrm{GF}(p)$: the Kroniker basis (i.e., $k_i(x) = 1$ if $x = i$ and $k_i(x) = 0$) and the (normalized) Fourier basis (i.e., $f_i(x) = p^{-1/2} \cdot \omega^{ix}$). Note that the biases of ζ correspond to the inner products of δ with the non-constant Fourier functions, whereas the distances of ζ from the uniform distribution correspond to the inner products of δ with the Kroniker functions.

Exercise 5.17 (other pseudorandom generators and the hitting problem) Show that various pseudorandom generators yield solutions to the hitting problem (as defined in Definition 5.4). Specifically:

1. Show that a pairwise independence generator of block-length b and stretch ℓ yields a sequence over $\{0,1\}^b$ that is (ε, δ)-hitting for $\delta = O(1/\varepsilon\ell')$, where $\ell' = \ell/b$.

   Advanced exercise: Show that when using t-wise independence. the error bound can be reduced to $\delta = O(t^2/\varepsilon\ell')^{\lfloor t/2 \rfloor}$.

2. Referring to Definition 4.1, show that a (b, δ)-pseudorandom generator of stretch ℓ yields a sequence over $\{0,1\}^b$ that is (ε, δ)-hitting for $\delta = (1-\varepsilon)^{\ell/b} + \delta$.

3. Consider modifications of the hitting problem in which the target set T is restricted to be recognizable within some specified complexity.

   (a) Show that a general-purpose pseudorandom generator of stretch ℓ yields a sequence over $\{0,1\}^b$ that is (ε, δ)-hitting for target sets in $\mathcal{BPP}$ and $\delta = (1-\varepsilon)^{\ell/b} + 1/p$, where p is an arbitrary polynomial.

   (b) Referring to Definition 3.1, show that a canonical derandomizer of stretch ℓ yields a sequence over $\{0,1\}^b$ that is (ε, δ)-hitting for target sets that are recognized by circuits of size ℓ^2 and $\delta = (1-\varepsilon)^{\ell/b} + 1/6$.

What is the advantage of using the expander random walk generator over each of the foregoing options?

Exercise 5.18 (a version of the Expander Random Walk Theorem) Let $G = (V, E)$ be a graph as in Theorem 5.5. Prove that the probability that a random walk of length ℓ' intersects $W_0 \times W_1 \times \cdots \times W_{\ell'-1} \subseteq V^{\ell'}$ is upper bounded by Eq. (5.8).

Guideline: Let A be a matrix representing the random walk on G (i.e., A is the adjacency matrix of G divided by d), and let $\hat{\lambda} \stackrel{\text{def}}{=} \lambda/d$. Note that the uniform distribution, represented by the vector $\overline{u} = (N^{-1}, ..., N^{-1})^\top$, is the eigenvector of A that is associated with the largest eigenvalue (which is 1), whereas all other eigenvalues have absolute value at most $\hat{\lambda}$. Let P_i be a 0-1 matrix that has 1-entries only on its diagonal such that entry (j,j) is set to 1 if and only if $j \in W_i$. Then, the probability that a random walk of length ℓ intersects $W_0 \times W_1 \times \cdots \times W_{\ell-1}$ is the sum of the entries of the vector $\overline{v} \stackrel{\text{def}}{=} P_{\ell-1}A \cdots P_2AP_1AP_0\overline{u}$. We are interested in upper-bounding $\|\overline{v}\|_1$, and use $\|\overline{v}\|_1 \leq \sqrt{N} \cdot \|\overline{v}\|$, where $\|\overline{z}\|_1$ and $\|\overline{z}\|$ denote the L_1-norm and L_2-norm of $\overline{z}$, respectively (e.g., $\|\overline{u}\|_1 = 1$ and $\|\overline{u}\| = N^{-1/2}$). The key observation is that the linear transformation P_iA shrinks every vector. For further details, see [24, Apdx. E.2.1.3].

Exercise 5.19 Using notation as in Theorem 5.5, prove that the probability that a random walk of length ℓ' visits W more than $\alpha\ell'$ times is smaller than $\binom{\ell'}{\alpha\ell'} \cdot (\rho + (\lambda/d)^2)^{\alpha\ell'/2}$. For example, for $\alpha = 1/2$ and $\lambda/d < \sqrt{\rho}$, we get an upper-bound of $(32\rho)^{\ell'/4}$. We comment that much better bounds can be obtained (cf., e.g., [33]).

Guideline: Use a union bound on all possible sequences of $m = \alpha\ell'$ visits, and upper-bound the probability of visiting W in steps $j_1, ..., j_m$ by applying Eq. (5.8) with $W_i = W$ if $i \in \{j_1, ..., j_m\}$ and $W = V$ otherwise.

Concluding Remarks

We discussed a variety of incarnations of the generic notion of a pseudorandom generator, leading to vastly different concrete notions of pseudorandom generators. Some of the latter notions are depicted in the following figure.

TYPE	distinguisher's resources	generator's resources	stretch (i.e., $\ell(k)$)	comments
gen.-purpose	$p(k)$-time, $\forall$ poly. p	poly(k)-time	poly(k)	Assumes OW
canon. derand.	$2^{k/O(1)}$-time	$2^{O(k)}$-time	$2^{k/O(1)}$	Assumes EvC
space-bounded	$s(k)$-space, $s(k) < k$	$O(k)$-space	$2^{k/O(s(k))}$	runs in time
robustness	$k/O(1)$-space	$O(k)$-space	poly(k)	poly$(k) \cdot \ell(k)$
t-wise indepen.	inspect t positions	poly$(k) \cdot \ell(k)$-time	$2^{k/O(t)}$	(e.g., pairwise)
small bias	linear tests	poly$(k) \cdot \ell(k)$-time	$2^{k/O(1)} \cdot \varepsilon(k)$	
expander	"hitting"	poly$(k) \cdot \ell(k)$-time	$\ell'(k) \cdot b(k)$	
random walk	$(0.5, 2^{-\Omega(\ell'(k))})$-hitting for $\{0,1\}^{b(k)}$, with $\ell'(k) = \Omega(k - b(k)) + 1$.			

By OW we denote the assumption that one-way functions exists, and by EvC we denote the assumption that the class $\mathcal{E}$ has (almost-everywhere) exponential circuit complexity.

Pseudorandom generators at a glance.

We highlight a key distinction between the case of general-purpose pseudorandom generators (treated in Chapter 2) and the other cases (cf. e.g., Chapters 3 and 4): in the former case the distinguisher is more complex than the generator, whereas in the latter cases the generator is more complex than the distinguisher. Specifically, a general-purpose generator runs in (some *fixed*) polynomial-time and needs to withstand *any* probabilistic polynomial-time distinguisher. In fact, some of the proofs presented in Chapter 2 utilize the fact that the distinguisher can invoke the generator on seeds of its choice. In contrast, the Nisan-Wigderson Generator, analyzed in Theorem 3.5, runs more time than the distinguishers that it tries to fool, and the proof relies on this fact in an essential manner. Similarly, the space-complexity of the space-resilient generators presented in Chapter 4 is higher than the space-bound of the distinguishers that they fool.

Reiterating some of the notes of Chapter 1, we stress that our presentation, which views vastly different notions of pseudorandom generators as incarnations of a general paradigm, has emerged mostly in retrospect. Nevertheless, while the historical study of the various notions was mostly unrelated at a technical level, the case of general-purpose pseudorandom generators served as a source of inspiration to most of the other cases. In particular, the concept of computational indistinguishability, the connection between hardness and pseudorandomness, and the equivalence between pseudorandomness and unpredictability, appeared first in the context of general-purpose pseudorandom generators (and inspired the development of "generators for derandomization" and "generators for space bounded machines").

We stress that the chapters' notes do not mention several technical contributions that played an important role in the development of the area. For further details, the interested reader is referred to [21, Chap. 3].

Finally, we mention that the study of pseudorandom generators is part of complexity theory, and the interested reader is encouraged to further explore the connections between pseudorandomness and complexity theory at large (cf. e.g., [24]).

Appendix A

Hashing Functions

Hashing is extensively used in computer science, where the typical application is for mapping arbitrary (unstructured) sets into a structured set of comparable size such that the mapping is "almost uniform". Specifically, hashing is used for mapping an arbitrary 2^m-subset of $\{0,1\}^n$ to $\{0,1\}^m$ in an "almost uniform" manner.

For any fixed set S of cardinality 2^m, there exists a one-to-one mapping $f_S : S \to \{0,1\}^m$, but this mapping is not necessarily efficiently computable (e.g., it may require "knowing" the entire set S). On the other hand, no single function $f : \{0,1\}^n \to \{0,1\}^m$ can map every 2^m-subset of $\{0,1\}^n$ to $\{0,1\}^m$ in a one-to-one manner (or even approximately so). Nevertheless, for every 2^m-subset $S \subset \{0,1\}^n$, a random function $f : \{0,1\}^n \to \{0,1\}^m$ has the property that, with overwhelmingly high probability, f maps S to $\{0,1\}^m$ such that no point in the range has too many f-preimages in S. The problem is that a truly random function is unlikely to have a succinct representation (let alone an efficient evaluation algorithm). We thus seek families of functions that have a "random mapping" property (as in Item 1 of the following definition), but do have a succinct representation as well as an efficient evaluation algorithm (as in Items 2 and 3 of the following definition).

A.1 Definitions

Motivated by the foregoing discussion, we consider families of functions $\{H_n^m\}_{m<n}$ such that the following properties hold:

1. For every $S \subset \{0,1\}^n$, with high probability, a function h selected uniformly in H_n^m maps S to $\{0,1\}^m$ in an "almost uniform" manner. For example, we may require that, for any $|S| = 2^m$ and each point y, with high probability over the choice of h, it holds that $|\{x \in S : h(x) = y\}| \leq \text{poly}(n)$.

2. The functions in H_n^m have succinct representation. For example, we may require that $H_n^m \equiv \{0,1\}^{\ell(n,m)}$, for some polynomial ℓ.

3. The functions in H_n^m can be efficiently evaluated. That is, there exists a polynomial-time algorithm that, on input a representation of a function, h (in H_n^m), and a string $x \in \{0,1\}^n$, returns $h(x)$. In some cases we make even more stringent requirements regarding the algorithm (e.g., that it runs in linear space).

Condition 1 was left vague on purpose. At the very least, we require that the expected size of $\{x \in S : h(x) = y\}$ equals $|S|/2^m$. We shall see (in Section A.3) that different interpretations of Condition 1 are satisfied by different families of hashing functions. We focus on t-wise independent hashing functions, defined next.

Definition A.1 (t-wise independent hashing functions): *A family H_n^m of functions from n-bit strings to m-bit strings is called* t-wise independent *if for every t distinct domain elements $x_1, ..., x_t \in \{0,1\}^n$ and every $y_1, ..., y_t \in \{0,1\}^m$ it holds that*

$$\mathsf{Pr}_{h \in H_n^m}\left[\bigwedge_{i=1}^{t} h(x_i) = y_i\right] = 2^{-t \cdot m}$$

That is, a uniformly chosen $h \in H_n^m$ maps every t domain elements to the range in a totally uniform manner. Note that for $t \geq 2$, it follows that the probability that a random $h \in H_n^m$ maps two distinct domain elements to the same image equals 2^{-m}. Such (families of) functions are called universal (cf. [13]), but we will focus on the stronger condition of t-wise independence.

A.2 Constructions

The following constructions are merely a re-interpretation of the constructions presented in Section 5.1.1. (Alternatively, one may view the constructions presented in Section 5.1.1 as a re-interpretation of the following two constructions.)

Construction A.2 (t-wise independent hashing): *For $t, m, n \in \mathbb{N}$ such that $m \leq n$, consider the following family of hashing functions mapping n-bit strings to m-bit strings. Each t-sequence $\overline{s} = (s_0, s_1, ..., s_{t-1}) \in \{0,1\}^{t \cdot n}$ describes a function $h_{\overline{s}} : \{0,1\}^n \to \{0,1\}^m$ such that $h_{\overline{s}}(x)$ equals the m-bit prefix of the binary representation of $\sum_{j=0}^{t-1} s_j x^j$, where the arithmetic is that of* $\mathrm{GF}(2^n)$, *the finite field of* 2^n *elements.*

Proposition 5.1 implies that Construction A.2 constitutes a family of t-wise independent hash functions. Typically, we will use either $t = 2$ or $t = \Theta(n)$. To make the construction totally explicit, we need an explicit representation of $\mathrm{GF}(2^n)$; see the comment following Proposition 5.1. An alternative construction for the case of $t = 2$ may be obtained analogously to the pairwise independent generator of Proposition 5.2. This construction, presented next, relies on Toeplitz matrices, where $T = (t_{i,j})$ is a Toeplitz matrix if $t_{i,j} = t_{i+1,j+1}$, for all i, j.

Construction A.3 (alternative pairwise independent hashing): *For $m \leq n$, consider the family of hashing functions in which each pair (T, b), consisting of an n-by-m Toeplitz matrix T and an m-dimensional vector b, describes a function $h_{T,b} : \{0,1\}^n \to \{0,1\}^m$ such that $h_{T,b}(x) = Tx + b$.*

Proposition 5.2 implies that Construction A.3 constitutes a family of pairwise independent hash functions. Note that an n-by-m Toeplitz matrix can be specified by $n + m - 1$ bits, yielding a description length of $n + 2m - 1$ bits. An alternative construction (analogous to Eq. (5.10) and requiring $m \cdot n + m$ bits of representation) uses arbitrary n-by-m matrices rather than Toeplitz matrices.

A.3 The Leftover Hash Lemma

We now turn to the "almost uniform" cover condition (i.e., Condition 1) mentioned in Section A.1. One concrete interpretation of this condition is given by the following lemma (and another interpretation is implied by it: see Theorem A.5).

Lemma A.4 (uniform cover, per each range element): *Let $m \leq n$ be integers, H_n^m be a family of pairwise independent hash functions, and $S \subseteq \{0,1\}^n$. Then, for every $y \in \{0,1\}^m$ and every $\varepsilon > 0$, for all but at most a $\frac{2^m}{\varepsilon^2 |S|}$ fraction of $h \in H_n^m$ it holds that*

$$(1-\varepsilon) \cdot \frac{|S|}{2^m} \;<\; |\{x \in S : h(x) = y\}| \;<\; (1+\varepsilon) \cdot \frac{|S|}{2^m}. \tag{A.1}$$

Note that by pairwise independence (or rather even by 1-wise independence), the expected size of $\{x \in S : h(x) = y\}$ is $|S|/2^m$, where the expectation is taken uniformly over all $h \in H_n^m$. The lemma upper bounds the fraction of h's that deviate from the expected behavior (i.e., for which $|h^{-1}(y) \cap S| \neq (1 \pm \varepsilon) \cdot |S|/2^m$). Needless to say, the bound is meaningful only when $|S| > 2^m/\varepsilon^2$. Focusing on the case that $|S| > 2^m$ and setting $\varepsilon = \sqrt[3]{2^m/|S|}$, we infer that *for all but at most an ε fraction of $h \in H_n^m$ it holds that $|\{x \in S : h(x) = y\}| = (1 \pm \varepsilon) \cdot |S|/2^m$*. Thus, each range element has approximately the right number of h-preimages in the set S, under almost all $h \in H_n^m$.

Proof: Fixing an arbitrary set $S \subseteq \{0,1\}^n$ and an arbitrary $y \in \{0,1\}^m$, we estimate the probability that a uniformly selected $h \in H_n^m$ violates Eq. (A.1). We define random variables ζ_x, over the aforementioned probability space, such that $\zeta_x = \zeta_x(h)$ equal 1 if $h(x) = y$ and $\zeta_x = 0$ otherwise. The expected value of $\sum_{x \in S} \zeta_x$ is $\mu \stackrel{\text{def}}{=} |S| \cdot 2^{-m}$, and we are interested in the probability that this sum deviates from the expectation. Applying Chebyshev's Inequality, we get

$$\mathsf{Pr}\left[\left|\mu - \sum_{x \in S} \zeta_x\right| \geq \varepsilon \cdot \mu\right] \;<\; \frac{\mu}{\varepsilon^2 \mu^2}$$

because $\mathsf{Var}[\sum_{x \in S} \zeta_x] < |S| \cdot 2^{-m}$ by the pairwise independence of the ζ_x's and the fact that $\mathsf{E}[\zeta_x] = 2^{-m}$. The lemma follows. ■

A generalization (called mixing). The proof of Lemma A.4 can be easily extended to show that *for every set $T \subset \{0,1\}^m$ and every $\varepsilon > 0$, for all but at most a $\frac{2^m}{|T| \cdot |S| \varepsilon^2}$ fraction of $h \in H_n^m$ it holds that $|\{x \in S : h(x) \in T\}| = (1 \pm \varepsilon) \cdot |T| \cdot |S|/2^m$.* (Hint: redefine $\zeta_x = \zeta(h) = 1$ if $h(x) \in T$ and $\zeta_x = 0$ otherwise.) This assertion is meaningful provided that $|T| \cdot |S| > 2^m/\varepsilon^2$, and in the case that $m = n$ it is called a mixing property.

A useful corollary. The aforementioned generalization of Lemma A.4 asserts that, for any fixed set of preimages $S \subset \{0,1\}^n$ and any fixed sets of images $T \subset \{0,1\}^m$, most functions in H_n^m behave well with respect to S and T (in the sense that they map approximately the adequate fraction of S (i.e., $|T|/2^m$) to T). A seemingly stronger statement, which is implied by Lemma A.4 itself, reverses the order of quantification with respect to T; that is, for all adequate sets S, most functions in H_n^m map S

to $\{0,1\}^m$ in an almost uniform manner (i.e., assign each set T approximately the adequate fraction of S, where here the approximation is up to an additive deviation). As we shall see, this is a consequence of the following theorem.

Theorem A.5 (a.k.a. Leftover Hash Lemma): *Let H_n^m and $S \subseteq \{0,1\}^n$ be as in Lemma A.4, and define $\varepsilon = \sqrt[3]{2^m/|S|}$. Consider random variables X and H that are uniformly distributed on S and H_n^m, respectively. Then, the statistical distance between $(H, H(X))$ and (H, U_m) is at most 2ε.*

It follows that, *for X and ε as in Theorem A.5 and any $\alpha > 0$, for all but at most an α fraction of the functions $h \in H_n^m$ it holds that $h(X)$ is $(2\varepsilon/\alpha)$-close to U_m.* (Using the terminology of the subsequent Section B.1, we may say that Theorem A.5 asserts that H_n^m yields a strong extractor.) The proof of Theorem A.5 is omitted, and the interested reader is referred to [24, Apdx. D.2.3].

Appendix B

On Randomness Extractors

Extracting almost-perfect randomness from sources of weak (i.e., defected) randomness is crucial for the actual use of randomized algorithms, procedures and protocols. The latter are analyzed assuming that they are given access to a perfect random source, while in reality one typically has access only to sources of weak (i.e., highly imperfect) randomness. This gap is bridged by using randomness extractors, which are efficient procedures that (possibly with the help of little extra randomness) convert any source of weak randomness into an almost-perfect random source. Thus, randomness extractors are devices that greatly enhance the quality of random sources. In addition, randomness extractors are related to several other fundamental problems (see, e.g., [24, Apdx. D.4.1] and [62]).

One key parameter, which was avoided in the foregoing abstract discussion, is the class of weak random sources from which we need to extract almost perfect randomness. Needless to say, it is preferable to make as little assumptions as possible regarding the weak random source. In other words, we wish to consider a wide class of such sources, and require that the randomness extractor (often referred to as the extractor) "works well" for any source in this class. A general class of such sources is defined in Section B.1, but first we wish to mention that even for very restricted classes of sources no deterministic extractor can work.[1] To overcome this impossibility result, two approaches are used:

Seeded extractors: The first approach consists of considering randomized extractors that use a relatively small amount of randomness (in addition to the weak random source). That is, these extractors obtain two inputs: a short truly random seed and a relatively long sequence generated by an arbitrary source that belongs to the specified class of sources. This suggestion is motivated in two different ways:

1. The application may actually have access to an almost-perfect random source, but bits from this high-quality source are much more expensive than bits from the weak (i.e., low-quality) random source. Thus, it makes sense to obtain a few high-quality bits from the almost-perfect source and use them to "purify" the cheap bits obtained from the weak (low-quality) source. Thus, combining

[1] For example, consider the class of sources that output n-bit strings such that no string occurs with probability greater than $2^{-(n-1)}$ (i.e., twice its probability weight under the uniform distribution).

many cheap (but low-quality) bits with few high-quality (but expensive) bits, we obtain many high-quality bits.

2. In some applications (e.g., when using randomized algorithms), it may be possible to invoke the application multiple times, and use the "typical" outcome of these invocations (e.g., rule by majority in the case of a decision procedure). For such applications, we may proceed as follows: First we obtain an outcome r of the weak random source, then we invoke the application multiple times such that for every possible seed s we invoke the application feeding it with $\mathtt{extract}(s, r)$, and finally we use the "typical" outcome of these invocations. Indeed, this is analogous to the context of derandomization (see Section 3), and likewise this alternative is typically not applicable to cryptographic and/or distributed settings.

Extraction from a few independent sources: The second approach consists of considering deterministic extractors that obtain samples from a few (say two) *independent* sources of weak randomness. Such extractors are applicable in any setting (including in cryptography), provided that the application has access to the required number of independent weak random sources.

In this appendix we focus on the first type of extractors (i.e., the *seeded extractors*). This choice is motivated by the applications in the main text as well by the closer connection between seeded extractors and other topics in complexity theory. We also mention that our understanding of seeded extractors seem much more mature than the current state of knowledge regarding extraction from a few independent sources. Below we only present a definition that corresponds to the foregoing motivational discussion, and mention that its relation to other topics in complexity theory is discussed in [24, Apdx. D.4.1] and in [62].

B.1 Definitions

A very wide class of weak random sources corresponds to sources in which no specific output is too probable. That is, the class is parameterized by a (probability) bound β and consists of all sources X such that for every x it holds that $\mathsf{Pr}[X = x] \leq \beta$. In such a case, we say that X has min-entropy[2] at least $\log_2(1/\beta)$. Indeed, we represent sources as random variables, and assume that they are distributed over strings of a fixed length, denoted n. An (n, k)-source is a source that is distributed over $\{0, 1\}^n$ and has min-entropy at least k.

An interesting special case of (n, k)-sources is that of sources that are uniform over some subset of 2^k strings. Such sources are called (n, k)-flat. A useful observation is that *each (n, k)-source is a convex combination of (n, k)-flat sources.*

Definition B.1 (extractor for (n, k)-sources):

1. *An algorithm* $\mathrm{Ext}: \{0,1\}^n \times \{0,1\}^d \to \{0,1\}^m$ *is called an* extractor with error ε for the class $\mathcal{C}$ *if for every source X in $\mathcal{C}$ it holds that* $\mathrm{Ext}(X, U_d)$ *is ε-close to U_m. If $\mathcal{C}$ is the class of (n, k)-sources, then* Ext *is called a* (k, ε)-extractor.

[2]Recall that the entropy of a random variable X is defined as $\sum_x \mathsf{Pr}[X = x] \cdot \log_2(1/\mathsf{Pr}[X = x])$. Indeed the min-entropy of X equals $\min_x\{\log_2(1/\mathsf{Pr}[X = x])\}$, and is always upper-bounded by its entropy.

2. *An algorithm* Ext *is called a* strong extractor with error ε for $\mathcal{C}$ *if for every source* X *in* $\mathcal{C}$ *it holds that* $(U_d, \text{Ext}(X, U_d))$ *is* ε*-close to* (U_d, U_m)*. A* strong (k, ε)-extractor *is defined analogously.*

Using the aforementioned "decomposition" of (n, k)-sources into (n, k)-flat sources, it follows that Ext *is a* (k, ε)*-extractor if and only if it is an extractor with error* ε *for the class of* (n, k)*-flat sources.* (A similar claim holds for strong extractors.) Thus, much of the technical analysis is conducted with respect to the class of (n, k)-flat sources. For example, by analyzing the case of (n, k)-flat sources it is easy to see that, for $d = \log_2(n/\varepsilon^2) + O(1)$, there exists a (k, ε)-extractor $\text{Ext} : \{0,1\}^n \times \{0,1\}^d \to \{0,1\}^k$. (The proof employs the Probabilistic Method and uses a union bound on the (finite) set of all (n, k)-flat sources.)[3]

We seek, however, explicit extractors; that is, extractors that are implementable by polynomial-time algorithms. We note that the evaluation algorithm of any family of pairwise independent hash functions mapping n-bit strings to m-bit strings constitutes a (strong) (k, ε)-extractor for $\varepsilon = 2^{-\Omega(k-m)}$ (see Theorem A.5). However, these extractors necessarily use a long seed (i.e., $d \geq 2m$ must hold (and in fact $d = n + 2m - 1$ holds in Construction A.3)). In Section B.2 we survey constructions of efficient (k, ε)-extractors that obtain logarithmic seed length (i.e., $d = O(\log(n/\varepsilon))$).

On the importance of logarithmic seed length. The case of logarithmic seed length (i.e., $d = O(\log(n/\varepsilon))$) is of particular importance for a variety of reasons. First, when emulating a randomized algorithm using a defected random source (as in Item 2 of the motivational discussion of seeded extractors), the overhead is exponential in the length of the seed. Thus, the emulation of a generic probabilistic polynomial-time algorithm can be done in polynomial time only if the seed length is logarithmic. Similar considerations apply to other applications of extractors. Last, we note that logarithmic seed length is an absolute lower-bound for (k, ε)-extractors, whenever $k < n - n^{\Omega(1)}$ (and the extractor is non-trivial (i.e., $m \geq 1$ and $\varepsilon < 1/2$)).

B.2 Constructions

Recall that we seek explicit constructions of extractors; that is, functions $\text{Ext} : \{0,1\}^n \times \{0,1\}^d \to \{0,1\}^m$ that can be computed in polynomial-time. The question, of course, is of parameters; that is, having explicit (k, ε)-extractors *with* m *as large as possible and* d *as small as possible.* We first note that, except for "pathological" cases[4], both $m \leq k + d - (2\log_2(1/\varepsilon) - O(1))$ and $d \geq \log_2((n-k)/\varepsilon^2) - O(1)$ must hold, regardless of the explicitness requirement. The aforementioned bounds are in fact tight; that is, there exist (non-explicit) (k, ε)-extractors with $m = k + d - 2\log_2(1/\varepsilon) - O(1)$ and $d = \log_2((n-k)/\varepsilon^2) + O(1)$. The obvious goal is meeting these bounds via explicit constructions.

---

[3]Indeed, the key fact is that the number of (n, k)-flat sources is $N \stackrel{\text{def}}{=} \binom{2^n}{2^k}$. The probability that a random function $\text{Ext} : \{0,1\}^n \times \{0,1\}^d \to \{0,1\}^k$ is not an extractor with error ε for a fixed (n, k)-flat source is upper-bounded by $p \stackrel{\text{def}}{=} 2^{2^k} \cdot \exp(-\Omega(2^{d+k}\varepsilon^2))$, because p bounds the probability that when selecting 2^{d+k} random k-bit long strings there exists a set $T \subset \{0,1\}^k$ that is hit by more than $((|T|/2^k) + \varepsilon) \cdot 2^{d+k}$ of these strings. Note that for $d = \log_2(n/\varepsilon^2) + O(1)$ it holds that $N \cdot p \ll 1$. In fact, the same analysis applies to the extraction of $m = k + \log_2 n$ bits (rather than k bits).

[4]That is, for $\varepsilon < 1/2$ and $m > d$.

Some known results. Despite tremendous progress on this problem (and occasional claims regarding "optimal" explicit constructions), the ultimate goal has not yet been reached. Nevertheless, the known explicit constructions are pretty close to being optimal.

Theorem B.2 (explicit constructions of extractors): *Explicit* (k, ε)*-extractors of the form* $\mathrm{Ext}: \{0,1\}^n \times \{0,1\}^d \to \{0,1\}^m$ *exist for the following cases (i.e., settings of the parameters* d *and* m*):*

1. *For* $d = O(\log n/\varepsilon)$ *and* $m = (1-\alpha)\cdot(k - O(d))$*, where* $\alpha > 0$ *is an arbitrarily small constant and provided that* $\varepsilon > \exp(-k^{1-\alpha})$.
2. *For* $d = (1+\alpha)\cdot\log_2 n$ *and* $m = k/\mathrm{poly}(\log n)$*, where* $\varepsilon, \alpha > 0$ *are arbitrarily small constants.*

Proofs of Part 1 and Part 2 can be found in [30] and [61], respectively. We note that, for the sake of simplicity, we did not quote the best possible bounds. Furthermore, we did not mention additional incomparable results (which are relevant for different ranges of parameters).

We refrain from providing an overview of the proof of Theorem B.2, but rather review the conceptual insight that underlies many of the results that belong to the current "generation" of constructions.

The pseudorandomness connection

The connection between extractors and certain pseudorandom generators, discovered by Trevisan [65], is the starting point of the current generation of constructions of extractors. This connection is surprising because it went in a non-standard direction; that is, transforming certain pseudorandom generators into extractors. We note that computational objects are typically more complex than the corresponding information theoretical objects (cf. e.g., Appendix C and [24, Chap. 7]). Thus, if pseudorandom generators and extractors are at all related (which was not suspected before [65]), then this relation should not be expected to help in the construction of extractors, which seem to be information theoretic objects. Nevertheless, the discovery of this relation did yield a breakthrough in the study of extractors.[5]

But before describing the connection, let us wonder for a moment. Just looking at the syntax, we note that pseudorandom generators have a single input (i.e., the seed), while extractors have two inputs (i.e., the n-bit long source and the d-bit long seed). But taking a second look at the Nisan–Wigderson Generator (i.e., the combination of Construction 3.4 with an amplification of worst-case to average-case hardness), we note that this construction can be viewed as taking two inputs: a d-bit long seed and a "hard" predicate on d'-bit long strings (where $d' = \Omega(d)$).[6] Now, an appealing idea is to use the n-bit long source as a (truth-table) description of a (worst-case) hard predicate (which indeed means setting $n = 2^{d'}$). The key observation is that *even if the source is only weakly random, then it is likely to represent a predicate that is inapproximable* (as in the hypothesis of Theorem 3.5).

[5] We note that once the connection became better understood, influence started going in the "right" direction: from extractors to pseudorandom generators.

[6] Indeed, to fit the current context, we have modified some notation. In Construction 3.4 the length of the seed is denoted by k and the length of the input for the predicate is denoted by m.

Recall that the aforementioned construction is supposed to yield a pseudorandom generator whenever it starts with a hard predicate. In the current context, where there are no computational restrictions, pseudorandomness is supposed to hold against any (computationally unbounded) distinguisher, and thus here pseudorandomness means being statistically close to the uniform distribution (on strings of the adequate length, denoted ℓ). Intuitively, this makes sense only if the observed sequence is shorter than the amount of randomness in the source (and seed), which is indeed the case (i.e., $\ell < k + d$, where k denotes the min-entropy of the source). Hence, there is hope to obtain a good extractor this way.

To turn the hope into reality, we need a proof (which is sketched next). Looking again at the Nisan–Wigderson Generator, we note that the proof of indistinguishability of this generator provides a black-box procedure for approximating the underlying predicate when given oracle access to any potential distinguisher. Specifically, in the proofs of Theorem 3.5 (which holds for any $\ell = 2^{\Omega(d')}$)[7], this black-box procedure was implemented by a *relatively small circuit* (which depends on the underlying predicate). Hence, this procedure contains relatively little information (regarding the underlying predicate), on top of the observed ℓ-bit long output of the extractor/generator. Specifically, for some fixed polynomial p, the amount of information encoded in the procedure (and thus available to it) is upper-bounded by $p(\ell)$, while the procedure is supposed to approximate the underlying predicate in the sense that this approximation determines a set of at most $p(\ell)$ predicates that contain the original predicate. Thus, $b = p(\ell)^2$ bits of information are supposed to fully determine the underlying predicate, which in turn is identical to the n-bit long source. However, if the source has min-entropy exceeding b, then it cannot be fully determined using only b bits of information.

It follows that the foregoing construction constitutes a $(b + O(1), 1/6)$-extractor (outputting $\ell = b^{\Omega(1)}$ bits), where the constant $1/6$ is the one used in the proof of Theorem 3.5 (and the argument holds provided that $b = n^{\Omega(1)}$). Note that this extractor uses a seed of length $d = O(d') = O(\log n)$. The argument can be extended to obtain $(k, \text{poly}(1/k))$-extractors that output $k^{\Omega(1)}$ bits using seeds of length $d = O(\log n)$, provided that $k = n^{\Omega(1)}$.

We stress that the foregoing description has only referred to two abstract properties of the Nisan–Wigderson Generator: (1) the fact that this generator uses any worst-case hard predicate as a black-box, and (2) the fact that its analysis uses any distinguisher as a black-box.

[7] Recalling that $n = 2^{d'}$, the restriction $\ell = 2^{\Omega(d')}$ implies $\ell = n^{\Omega(1)}$.

Appendix C

A Generic Hard-Core Predicate

In this appendix, we provide a proof of Theorem 2.11. This is done because, in our opinion, at the last account, the conversion of computational difficulty to pseudorandomness occurs in this result. On the other hand, the proof of Theorem 2.11 is too long to fit to the main text without damaging the main thread of the presentation.

We mention that Theorem 2.11 may also be viewed as a "hardness amplification" result. For further details and related "hardness amplification" results, the interested reader is referred to [24, Chap. 7].

The basic strategy. The proof of Theorem 2.11 proceeds by a so-called reducibility argument, which is actually a reduction, but one that is analyzed with respect to average case complexity. Specifically, we reduce the task of inverting f to the task of predicting the hard-core of f', while making sure that the reduction (when applied to input distributed as in the inverting task) generates a distribution as in the definition of the predicting task. Thus, a contradiction to the claim that b is a hard-core of f' yields a contradiction to the hypothesis that f is hard to invert. We stress that this argument is far more complex than analyzing the corresponding "probabilistic" situation (i.e., the distribution of $(r, b(X,r))$, where $r \in \{0,1\}^n$ is uniformly distributed and X is a random variable with super-logarithmic min-entropy (which represents the "effective" knowledge of x, when given $f(x)$).[1]

Our starting point is a probabilistic polynomial-time algorithm A' that satisfies, for some polynomial p and infinitely many n's, $\mathsf{Pr}[A'(f(X_n), U_n) = b(X_n, U_n)] > (1/2) + (1/p(n))$, where X_n and U_n are uniformly and independently distributed over $\{0,1\}^n$. Using a simple averaging argument, we focus on an $\varepsilon \stackrel{\text{def}}{=} 1/2p(n)$ fraction of the x's for which $\mathsf{Pr}[A'(f(x), U_n) = b(x, U_n)] > (1/2) + \varepsilon$ holds. We will show how to use A' in order to invert f, on input $f(x)$, provided that x is in this good set (which has density ε). The crux of the entire proof is thus captured by the following result.

---

[1]The min-entropy of X is defined as $\min_v\{\log_2(1/\mathsf{Pr}[X = v])\}$; that is, if X has min-entropy m, then $\max_v\{\mathsf{Pr}[X = v]\} = 2^{-m}$. The Leftover Hashing Lemma (see Appendix A) implies that, in this case, $\mathsf{Pr}[b(X, U_n) = 1|U_n] = \frac{1}{2} \pm 2^{-\Omega(m)}$, where U_n denotes the uniform distribution over $\{0,1\}^n$.

Theorem C.1 (Theorem 2.11, revisited): *There exists a probabilistic oracle machine that, given parameters n, ε and oracle access to any function $B : \{0,1\}^n \to \{0,1\}$, halts after* poly$(n/\varepsilon)$ *steps and with probability at least 1/2 outputs a list of all strings $x \in \{0,1\}^n$ that satisfy*

$$\mathsf{Pr}_{r\in\{0,1\}^n}[B(r) = b(x,r)] \geq \frac{1}{2} + \varepsilon, \tag{C.1}$$

where $b(x,r)$ denotes the inner-product mod 2 of x and r.

This machine can be modified such that, with high probability, its output list does not include any string x such that $\mathsf{Pr}_{r\in\{0,1\}^n}[B(r) = b(x,r)] < \frac{1}{2} + \frac{\varepsilon}{2}$. However, the point is that using the foregoing machine, we can obtain an f-preimage of $f(x)$, whenever x belongs to the good set (i.e., satisfies Eq. (C.1) with respect to $B(r) \stackrel{\text{def}}{=} A'(f(x), r)$). Indeed, Theorem 2.11 follows from Theorem C.1 by emulating an oracle $B = B_x$ such that the query r is answered with the value $A'(f(x), r)$. That is, on input $f(x)$, we invoke the oracle machine while emulating the oracle B, and when the oracle machine halts and provides a list of candidates we check whether this list contains a preimage of $f(x)$ under f and output such a preimage if found. (Alternatively, we may just output at random one of the candidates in the said list.)

Proof: It is instructive to think about any string x that satisfies Eq. (C.1).[2] We are given access to an oracle (or "black box") B that approximates $b(x,\cdot)$ with a non-negligible advantage over a coin toss; that is, $p_x \stackrel{\text{def}}{=} \mathsf{Pr}_{r\in\{0,1\}^n}[B(r) = b(x,r)]$ is at least $\frac{1}{2}+\varepsilon$ (as per Eq. (C.1)). Our task is to retrieve x, while making relatively few (i.e., poly(n/ε)-many) queries to B. Note that this would have been easy if B makes no errors at all (i.e., $p_x = 1$), but we face the case in which B's error rate is extremely high (i.e., it is only non-negligibly lower than the error rate of purely random noise). Also note that retrieving x based on 2^n queries to B is quite easy (also at a large error rate), but our goal is to operate in time that is inversely proportional to the advantage of B over a random coin toss.

A warm-up. Suppose for a moment that we replace the condition $p_x \geq \frac{1}{2} + \varepsilon$ by the much relaxed condition $p_x \geq \frac{3}{4} + \varepsilon$. In this case, retrieving x, by using B, is quite easy: To retrieve the ith bit of x, denoted x_i, we randomly select $r \in \{0,1\}^{|x|}$, and obtain $B(r)$ and $B(r \oplus e^i)$, where $e^i = 0^{i-1}10^{|x|-i}$ and $v \oplus u$ denotes the addition mod 2 of the binary vectors v and u. A key observation underlying the foregoing scheme as well as the rest of the proof is that $b(x, r \oplus s) = b(x,r) \oplus b(x,s)$, which can be readily verified by writing $b(x,y) = \sum_{i=1}^{n} x_i y_i \bmod 2$ and noting that addition modulo 2 of bits corresponds to their XOR. Now, note that if both $B(r) = b(x,r)$ and $B(r \oplus e^i) = b(x, r \oplus e^i)$ hold, then $B(r) \oplus B(r \oplus e^i)$ equals $b(x,r) \oplus b(x, r\oplus e^i) = b(x, e^i) = x_i$. The probability that both $B(r) = b(x,r)$ and $B(r \oplus e^i) = b(x, r \oplus e^i)$ hold, for a random r, is at least $1 - 2 \cdot (1 - p_x) \geq \frac{1}{2} + 2\varepsilon$. Hence, repeating the foregoing procedure sufficiently many times (using independent random choices of such r's) and ruling by majority, we retrieve x_i with very high probability. Similarly, we can retrieve all the bits of x. However, the entire analysis refers to retrieving x when $p_x \geq \frac{3}{4} + \varepsilon$ holds, whereas we need to retrieve x also if only $p_x \geq \frac{1}{2} + \varepsilon$ holds.

[2] We note that, in general, there may be $O(1/\varepsilon^2)$ strings that satisfy Eq. (C.1). We also note that there may be at most one string x such that $\mathsf{Pr}_r[B(r) = b(x,r)] > 3/4$ holds.

The "error-doubling" phenomenon. The problem with the foregoing procedure is that it doubles the original error probability of $B(\cdot)$ with respect to $b(x, \cdot)$. Under the unrealistic (foregoing) assumption that B's error rate is non-negligibly smaller than $\frac{1}{4}$, the "error-doubling" phenomenon poses no problems. However, in general (and even in the special case where B's error is exactly $\frac{1}{4}$) the foregoing procedure is unlikely to retrieve x. Note that the error rate of B cannot be decreased by repeating B several times (e.g., for every x, it may be that B always answers correctly on three quarters of the possible r's, and always errs on the remaining quarter). What is required is an *alternative way of using* B, a way that does not double the original error probability of B.

The key idea is generating the r's in a way that allows invoking B only once per each r (and i), instead of twice. Specifically, we will invoke B on $r \oplus e^i$ in order to obtain a "guess" for $b(x, r \oplus e^i)$, and obtain $b(x, r)$ in a different way (which does not involve using B). The good news is that the error probability is no longer doubled, since we only use B to get a "guess" of $b(x, r \oplus e^i)$. The bad news is that we still need to know $b(x, r)$, and it is not clear how we can know $b(x, r)$ without applying B. The answer is that we can guess $b(x, r)$ by ourselves. This is fine if we only need to guess $b(x, r)$ for one r (or logarithmically in $|x|$ many r's), but the problem is that we need to know (and hence guess) the value of $b(x, r)$ for polynomially many r's. The obvious way of guessing these $b(x, r)$'s yields an exponentially small success probability. Instead, we generate these polynomially many r's such that, on one hand they are "sufficiently random" whereas, on the other hand, we can guess all of the $b(x, r)$'s with noticeable success probability.[3] Specifically, generating the r's in a specific *pairwise independent* manner will satisfy both of these (conflicting) requirements. We stress that in case we are successful (in our guesses for all of the $b(x, r)$'s), we can retrieve x with high probability.

A word about the way in which the pairwise independent r's are generated (and the corresponding $b(x, r)$'s are guessed) is indeed in place. To generate $m = \text{poly}(|x|/\varepsilon)$ many r's, we uniformly (and independently) select $\ell \stackrel{\text{def}}{=} \log_2(m+1)$ strings in $\{0,1\}^{|x|}$. Let us denote these strings by $s^1, ..., s^\ell$. We then guess $b(x, s^1)$ through $b(x, s^\ell)$. Let us denote these guesses, which are uniformly (and independently) chosen in $\{0,1\}$, by σ^1 through σ^ℓ. Hence, the probability that all our guesses for the $b(x, s^i)$'s are correct is $2^{-\ell} = \frac{1}{\text{poly}(|x|)}$. The different r's correspond to the different *non-empty* subsets of $\{1, 2, ..., \ell\}$. Specifically, for every such subset J, we let $r^J \stackrel{\text{def}}{=} \bigoplus_{j \in J} s^j$. The reader can easily verify that the r^J's are pairwise independent and each is uniformly distributed in $\{0,1\}^{|x|}$; see Exercise 5.4. The key observation is that $b(x, r^J) = b(x, \bigoplus_{j \in J} s^j) = \bigoplus_{j \in J} b(x, s^j)$. Hence, our guess for $b(x, r^J)$ is $\bigoplus_{j \in J} \sigma^j$, and with noticeable probability all of our guesses are correct. Wrapping everything up, we obtain the following procedure, which makes oracle calls to B.

Retrieving procedure (accessing B, with parameters n and ε):
Set $\ell = \log_2(n/\varepsilon^2) + O(1)$.
(1) Select uniformly and independently $s^1, ..., s^\ell \in \{0,1\}^n$.
Select uniformly and independently $\sigma^1, ..., \sigma^\ell \in \{0,1\}$.
(2) For every non-empty $J \subseteq [\ell]$, compute $r^J \leftarrow \bigoplus_{j \in J} s^j$ and $\rho^J \leftarrow \bigoplus_{j \in J} \sigma^j$.

[3] Alternatively, we could try all polynomially many possible guesses, but our analysis does not benefit from this alternative.

(3) For $i = 1, ..., n$, determine the bit z_i according to the majority vote of the $(2^\ell - 1)$-long sequence of bits $(\rho^J \oplus B(r^J \oplus e^i))_{\emptyset \neq J \subseteq [\ell]}$.
(4) Output $z_1 \cdots z_n$.

Note that the "voting scheme" employed in Step 3 uses pairwise independent samples (i.e., the r^J's), but works essentially as well as it would have worked with independent samples (i.e., the independent r's).[4] That is, for every i and J, it holds that $\mathsf{Pr}_{s^1,...,s^\ell}[B(r^J \oplus e^i) = b(x, r^J \oplus e^i)] = p_x$ (which is at least $(1/2) + \varepsilon$), where $r^J = \bigoplus_{j \in J} s^j$, and (for every fixed i) the events corresponding to different J's are pairwise independent. It follows that *if for every $j \in [\ell]$ it holds that $\sigma^j = b(x, s^j)$*, then for every i and J we have

$$\begin{aligned} &\mathsf{Pr}_{s^1,...,s^\ell}[\rho^J \oplus B(r^J \oplus e^i) = b(x, e^i)] \qquad\qquad (C.2) \\ &= \mathsf{Pr}_{s^1,...,s^\ell}[B(r^J \oplus e^i) = b(x, r^J \oplus e^i)] > \frac{1}{2} + \varepsilon \end{aligned}$$

where the equality is due to $\rho^J = \bigoplus_{j \in J} \sigma^j = b(x, r^J) = b(x, r^J \oplus e^i) \oplus b(x, e^i)$. Note that Eq. (C.2) refers to the correctness of a single vote for $b(x, e^i)$. Using $m = 2^\ell - 1 = O(n/\varepsilon^2)$ and noting that these (Boolean) votes are pairwise independent, we infer that the probability that the majority of these votes is wrong is upper-bounded by $1/2n$. Using a union bound on all i's, we infer that with probability at least $1/2$, all majority votes are correct and thus x is retrieved correctly. Recall that the foregoing is conditioned on $\sigma^j = b(x, s^j)$ for every $j \in [\ell]$, which in turn holds with probability $2^{-\ell} = (m + 1)^{-1} = \Omega(\varepsilon^2/n)$. Thus, each x that satisfies Eq. (C.1) is retrieved correctly with probability $p \stackrel{\text{def}}{=} \Omega(\varepsilon^2/n)$.

Noting that x is merely a string for which Eq. (C.1) holds, it follows that the number of strings that satisfy Eq. (C.1) is at most $1/p$. Furthermore, by iterating the foregoing procedure for $\widetilde{O}(1/p)$ times we can obtain all of these strings. The theorem follows. ■

Digest. Theorem C.1 means that if given some information about x it is hard to recover x, then given the same information and a random r it is hard to predict $b(x, r)$. Indeed, the foregoing statement is in the spirit of Theorem 2.11 itself, except that it refers to any "information about x" (rather than to the value $f(x)$). To demonstrate the point, let us rephrase the foregoing statement as follows: *For every randomized process Π, if given s it is hard to obtain $\Pi(s)$, then given s and a uniformly distributed $r \in \{0,1\}^{|\Pi(s)|}$ it is hard to predict $b(\Pi(s), r)$.*

[4] Our focus here is on the accuracy of the approximation obtained by the sample, and not so much on the error probability. We wish to approximate $\mathsf{Pr}[b(x,r) \oplus B(r \oplus e^i) = 1]$ up to an additive term of ε, because such an approximation allows us to correctly determine $b(x, e^i)$. A pairwise independent sample of $O(t/\varepsilon^2)$ points allows for an approximation of a value in $[0, 1]$ up to an additive term of ε with error probability $1/t$, whereas a totally random sample of the same size yields error probability $\exp(-t)$. Since we can afford setting $t = \text{poly}(n)$ and having error probability $1/2n$, the difference in the error probability between the two approximation schemes is not important here.

Appendix D

Using Randomness in Computation

The underlying thesis of this primer is that randomness is playing an important role in computation. But since this primer is directed also at readers who are not closely familiar with the theory of computation, we feel that this thesis may require a short justification. Furthermore, our guess is that the proposition that there is a connection between computation and randomness may meet the skepticism of some readers, because computation seems the ultimate manifestation of determinism.

Still, a more sophisticated look at computation reveals that algorithms for solving standard search and decision problems as well as algorithmic strategies for multi-party interaction may benefit by using random choices. This is easiest to demonstrate in the domain of cryptography (see Appendix E) as well as in many other distributed and/or interactive settings (see, e.g., [8, 39, 40] and [24, Chap. 9], respectively). In this appendix, we consider the more basic setting of stand-alone computation, and present three simple randomized algorithms that solve basic computational problems. Many more examples can be found in [47].

D.1 A Simple Probabilistic Polynomial-Time Primality Test

Although a deterministic polynomial-time primality tester was found a few years ago [1], we believe that the following example provides a nice illustration to the power of randomized algorithms. We present a simple probabilistic polynomial-time algorithm for deciding whether or not a given number is a prime. The only Number Theoretic facts that we use are:

Fact 1: For every prime $p > 2$, each quadratic residue mod p has exactly two square roots mod p (and they sum up to p). That is, for every $r \in \{1, ..., p-1\}$, the equation $x^2 \equiv r^2 \pmod{p}$ has two solutions modulo p (i.e., r and $p - r$).

Fact 2: For every odd composite number N such that $N \neq M^e$ for all integers M and e, each quadratic residue mod N has at least four square roots mod N.

Our algorithm uses as a black-box an algorithm, denoted sqrt, that given a prime p and a quadratic residue mod p, denoted s, returns the smallest among the two modular square roots of s. There is no guarantee as to what the output is in the case that the input is not of the aforementioned form (and in particular in the case that p is not a prime). Thus, we actually present a probabilistic polynomial-time reduction of testing primality to extracting square roots modulo a prime (which is a search problem with a promise; see [24, Sec. 2.4.1]).

Construction D.1 (the reduction):[1] *On input a natural number $N > 2$, proceed as follows:*

1. *If N is either even or an integer-power,*[2] *then* reject.

2. *Uniformly select $r \in \{1, ..., N-1\}$, and set $s \leftarrow r^2 \bmod N$.*

3. *Let $r' \leftarrow \texttt{sqrt}(s, N)$. If $r' \equiv \pm r \pmod{N}$, then* accept *else* reject.

Indeed, in the case that N is composite, the reduction invokes sqrt on an illegitimate input (i.e., it makes a query that violates the promise of the problem at the target of the reduction). In such a case, there is no guarantee as to what sqrt answers, but actually a bluntly wrong answer only plays in our favor. In general, we will show that if N is a composite number, then the reduction rejects with probability at least $1/2$, regardless of how sqrt answers. We mention that there exists a probabilistic polynomial-time algorithm for implementing sqrt.

Proposition D.2 (analysis of the reduction): *Construction D.1 constitutes a probabilistic polynomial-time reduction of testing primality to extracting square roots module a prime. Furthermore, if the input is a prime, then the reduction always accepts, and otherwise it rejects with probability at least $1/2$.*

We stress that Proposition D.2 refers to the reduction itself; that is, sqrt is viewed as a ("perfect") oracle that, for every prime P and quadratic residue $s \pmod{P}$, returns $r < s/2$ such that $r^2 \equiv s \pmod{P}$. Combining Proposition D.2 with a probabilistic polynomial-time algorithm that computes sqrt with negligible error probability, we obtain that testing primality is in $\mathcal{BPP}$.

Proof: By Fact 1, on input a prime number N, Construction D.1 always accepts (because in this case, for every $r \in \{1, ..., N-1\}$, it holds that $\texttt{sqrt}(r^2 \bmod N, N) \in \{r, N-r\}$). On the other hand, suppose that N is an odd composite that is not an integer-power. Then, by Fact 2, each quadratic residue s has at least four square roots, and each of these square roots is equally likely to be chosen at Step 2 (in other words, s yields no information regarding which of its modular square roots was selected in Step 2). Thus, for every such s, the probability that either $\texttt{sqrt}(s, N)$ or $N - \texttt{sqrt}(s, N)$ equal the root chosen in Step 2 is at most $2/4$. It follows that, on input a composite number, the reduction rejects with probability at least $1/2$. ■

[1] Commonly attributed to Manuel Blum.

[2] This can be checked by scanning all possible powers $e \in \{2, ..., \log_2 N\}$, and (approximately) solving the equation $x^e = N$ for each value of e (i.e., finding the smallest integer i such that $i^e \geq N$). Such a solution can be found by a binary search.

Reflection: Construction D.1 illustrates an interesting aspect of randomized algorithms (or rather reductions); that is, their ability to take advantage of information that is unknown to the invoked subroutine. Specifically, Construction D.1 generates a problem instance (N, s), which hides crucial information (regarding how s was generated; i.e., which r such that $r^2 \equiv s \pmod N$ was selected in Step 2). Thus, $\mathtt{sqrt}(s, N)$ is oblivious of this hidden information (i.e., the identity of r), and so the quantity of interest is $\mathsf{Pr}_{r \in S_N(s)}[\mathtt{sqrt}(s, N) \in \{r, N - r\}]$, where $S_N(s)$ denotes the set of square roots of s modulo N.

Recall that testing primality is actually in $\mathcal{P}$. However, the deterministic algorithm demonstrating this fact is more complex than Construction D.1 (and its analysis is even more complicated).

D.2 Testing Polynomial Identity

An appealing example of a (one-sided error) randomized algorithm refers to the problem of determining whether two polynomials are identical. For simplicity, we assume that we are given an oracle for the evaluation of each of the two polynomials. An alternative presentation that refers to polynomials that are represented by arithmetic circuits yields a standard decision problem in co$\mathcal{RP}$ (the class of decision problems that are solvable by probabilistic polynomial-time algorithms that never reject a yes-instance).[3] Either way, we refer to multi-variant polynomials and to the question of whether they are identical over any field (or, equivalently, whether they are identical over a sufficiently large finite field). Note that it suffices to consider finite fields that are larger than the degree of the two polynomials.

Construction D.3 (Polynomial-Identity Test): *Let n be an integer and* F *be a finite field. Given black-box access to $p, q : \mathrm{F}^n \to \mathrm{F}$, uniformly select $r_1, ..., r_n \in \mathrm{F}$, and accept if and only if $p(r_1, ..., r_n) = q(r_1, ..., r_n)$.*

Clearly, if $p \equiv q$, then Construction D.3 always accepts. The following lemma implies that if p and q are different polynomials, each of total degree at most d over the finite field F, then Construction D.3 accepts with probability at most $d/|\mathrm{F}|$.

Lemma D.4 [60, 74]: *Let $p : \mathrm{F}^n \to \mathrm{F}$ be a non-zero polynomial of total degree d over the finite field* F*. Then*

$$\mathsf{Pr}_{r_1,...,r_n \in \mathrm{F}}[p(r_1, ..., r_n) = 0] \leq \frac{d}{|\mathrm{F}|}.$$

Proof: The lemma is proven by induction on n. The base case of $n = 1$ follows immediately by the Fundamental Theorem of Algebra (i.e., any non-zero univariate polynomial of degree d has at most d distinct roots). In the induction step, we write p as a polynomial in its first variable with coefficients that are polynomials in the other variables. That is,

$$p(x_1, x_2, ..., x_n) = \sum_{i=0}^{d} p_i(x_2, ..., x_n) \cdot x_1^i$$

[3] Equivalently, a set S is in co$\mathcal{RP}$ if and only if $\overline{S} \stackrel{\text{def}}{=} \{0,1\}^* \setminus S$ is in $\mathcal{RP}$.

where p_i is a polynomial of total degree at most $d-i$. Let j be the largest integer for which p_j is not identically zero. Dismissing the case $j = 0$ and using the induction hypothesis, we have

$$\begin{aligned} &\Pr_{r_1,r_2,...,r_n}[p(r_1,r_2,...,r_n)=0] \\ &\leq \Pr_{r_2,...,r_n}[p_j(r_2,...,r_n)=0] \\ &\quad + \Pr_{r_1,r_2,...,r_n}[p(r_1,r_2,...,r_n)=0 \,|\, p_j(r_2,...,r_n)\neq 0] \\ &\leq \frac{d-j}{|\mathrm{F}|}+\frac{j}{|\mathrm{F}|} \end{aligned}$$

where the second term is upper bounded by fixing any sequence $r_2, ..., r_n$ such that $p_j(r_2, ..., r_n) \neq 0$ and considering the univariate polynomial $p'(x) \stackrel{\text{def}}{=} p(x, r_2, ..., r_n)$ (which by hypothesis is a non-zero polynomial of degree j). ■

Reflection: Lemma D.4 may be viewed as asserting that for every non-zero polynomial of degree d over F at least a $1-(d/|\mathrm{F}|)$ fraction of its domain does not evaluate to zero. Thus, if $d \ll |\mathrm{F}|$, then most of the evaluation points constitute a witness for the fact that the polynomial is non-zero. We know of no efficient deterministic algorithm that, given a representation of the polynomial via an arithmetic circuit, finds such a witness. Indeed, Construction D.3 attempts to find a witness by merely selecting it at random.

D.3 The Accidental Tourist Sees It All

An appealing example of a randomized log-space algorithm is presented next. It refers to the problem of deciding undirected connectivity, and demonstrates that this problem is in $\mathcal{RL}$ (the log-space restriction of $\mathcal{RP}$). We mention that a deterministic log-space algorithm for this problem was found a few years ago (see [56]), but again the deterministic algorithm and its analysis are more complicated.

For the sake of simplicity, we consider the following computational problem: *Given an undirected graph G and a pair of vertices (s,t), determine whether or not s and t are connected in G.* Note that deciding undirected connectivity (of a given undirected graph) is log-space reducible to the foregoing problem (e.g., just check the connectivity of all pairs of vertices).

Construction D.5 (the random walk test): *On input (G, s, t), the randomized algorithm starts a* poly$(|G|)$*-long random walk at vertex s, and accepts the triple if and only if the walk passed through the vertex t. By a random walk we mean that at each step the algorithm selects uniformly one of the neighbors of the current vertex and moves to it.*

Observe that the algorithm can be implemented in logarithmic space (because we only need to store the current vertex as well as the number of steps taken so far). Obviously, if s and t are not connected in G, then the algorithm always rejects (G, s, t). Proposition D.6 implies that if s and t are connected (in G), then the algorithm accepts with probability at least $1/2$. It follows that undirected connectivity is in $\mathcal{RL}$.

Proposition D.6 [3]: *With probability at least* $1/2$*, a random walk of length* $O(|V| \cdot |E|)$ *starting at any vertex of the graph* $G = (V, E)$ *passes through all the vertices that reside in the same connected component as the start vertex.*

Thus, such a random walk may be used to explore the relevant connected component (in any graph). Following this walk one is likely to see all that there is to see in that component.

Proof Sketch: We will actually show that if G is connected, then, with probability at least $1/2$, a random walk starting at s visits all the vertices of G. For any pair of vertices (u, v), let $X_{u,v}$ be a random variable representing the number of steps taken in a random walk starting at u until v is *first encountered.* The reader may verify that for every edge $\{u, v\} \in E$ it holds that $\mathrm{E}[X_{u,v}] \leq 2|E|$. Next, we let $\mathrm{cover}(G)$ denote the expected number of steps in a random walk starting at s and ending when the last of the vertices of V is encountered. Our goal is to upper-bound $\mathrm{cover}(G)$. Towards this end, we consider an arbitrary directed cyclic-tour C that visits all vertices in G, and note that

$$\mathrm{cover}(G) \leq \sum_{(u,v) \in C} \mathrm{E}[X_{u,v}] \leq |C| \cdot 2|E|.$$

In particular, selecting C as a traversal of some spanning tree of G, we conclude that $\mathrm{cover}(G) < 4 \cdot |V| \cdot |E|$. Thus, with probability at least $1/2$, a random walk of length $8 \cdot |V| \cdot |E|$ starting at s visits all vertices of G. □

Appendix E

Cryptographic Applications of Pseudorandom Functions

A major application of random (or unpredictable) values is to the area of Cryptography. In fact, the very notion of a *secret* refers to such a random (or unpredictable) value. Furthermore, various natural security concerns (e.g., private communication) can be met by employing procedures that make essential use of such secrets and/or random values.

The extensive use of randomness in Cryptography makes this field a main client of pseudorandomness notions, techniques, and results. These are used not only in order to save on randomness (as in other algorithmic applications), but are rather essential to several basic cryptographic applications (see [23]).

In this appendix we focus on two major applications of *pseudorandom functions* to Cryptography; specifically, we use pseudorandom functions to construct schemes for providing secret and authenticated communication. That is, the two applications are secret communication and authenticated communication. In each of these cases, we first describe the application, and then describe how pseudorandom functions are used in order to achieve it. Detailed analysis of the two constructions can be found in [23, Sec. 5.3.3&6.3.1].

E.1 Secret Communication

The problem of providing *secret communication over insecure media* is the traditional and most basic problem of Cryptography. The setting of this problem consists of two parties communicating through a channel that is possibly tapped by an adversary. The parties wish to exchange information with each other, but keep the "wire-tapper" as ignorant as possible regarding the contents of this information. The canonical solution to the above problem is obtained by the use of encryption schemes.

Loosely speaking, an encryption scheme is a protocol allowing these parties to communicate *secretly* with each other. Typically, the encryption scheme consists of a pair of algorithms. One algorithm, called encryption, is applied by the sender (i.e., the party sending a message), while the other algorithm, called decryption, is applied by the receiver. Hence, in order to send a message, the sender first applies the encryption algorithm to the message, and sends the result, called the ciphertext, over the channel. Upon receiving a ciphertext, the other party (i.e., the receiver) applies the decryption algorithm to it, and retrieves the original message (called the plaintext).

In order for the foregoing scheme to provide secret communication, the communicating parties (at least the receiver) must know something that is not known to the wire-tapper. (Otherwise, the wire-tapper can decrypt the ciphertext exactly as done by the receiver.) This extra knowledge may take the form of the decryption algorithm itself, or some parameters and/or auxiliary inputs used by the decryption algorithm. We call this extra knowledge the decryption-key. Note that, without loss of generality, we may assume that the decryption algorithm is known to the wire-tapper, and that the decryption algorithm operates on two inputs: a ciphertext and a decryption-key. (The encryption algorithm also takes two inputs: a corresponding encryption-key and a plaintext.) We stress that the existence of a decryption-key, not known to the wire-tapper, is merely a necessary condition for secret communication.

The point we wish to make is that the decryption-key must be generated by a randomized algorithm. Suppose, in contrary, that the decryption-key is a predetermined function of publicly available data (i.e., the key is generated by employing an efficient deterministic algorithm to this data). Then, the wire-tapper can just obtain the key in exactly the same manner (i.e., invoking the same algorithm on the said data). We stress that saying that the wire-tapper does not know which algorithm to employ or does not have the data on which the algorithm is employed just shifts the problem elsewhere; that is, the question remains as to *how do the legitimate parties select this algorithm and/or the data to which it is applied*? Again, deterministically selecting these objects based on publicly available data will not do. At some point, *the legitimate parties must obtain some object that is unpredictable by the wire-tapper*, and such unpredictability refers to randomness (or pseudorandomness).

However, the role of randomness in allowing for secret communication is not confined to the generation of secret keys. To see why this is the case, we need to understand what "secrecy" is (i.e., to properly define what is meant by this intuitive term). Loosely speaking, we say that an encryption scheme is secure if it is *infeasible for the wire-tapper to obtain from the ciphertexts any additional information about the corresponding plaintexts.* In other words, whatever can be efficiently computed based on the ciphertexts can be efficiently computed from scratch (or rather from the a priori known data). Now, assuming that the encryption algorithm is deterministic, encrypting the same plaintext twice (using the same encryption-key) results in two identical ciphertexts, which are easily distinguishable from any pair of different ciphertexts resulting from the encryption of two different plaintexts. This problem does not arise when employing a randomized encryption algorithm (as presented next).

An encryption scheme based on pseudorandom functions. As indicated, an encryption scheme must also specify a method for selecting keys. In the following encryption scheme, the key is a uniformly selected n-bit string, denoted s. The parties use this key to determine a pseudorandom function f_s (as in Definition 2.17). A plaintext $x \in \{0,1\}^n$ is encrypted (using the key s) by uniformly selecting $r \in \{0,1\}^n$ and producing the ciphertext $(r, f_s(r) \oplus x)$, where $\alpha \oplus \beta$ denotes the bit-by-bit exclusive-or of the strings α and β. A ciphertext (r, y) is decrypted (using the key

s) by computing $f_s(r) \oplus y$. The security of this scheme follows from the security of an imaginary (ideal) scheme in which f_s is replaced by a totally random function $F : \{0,1\}^n \to \{0,1\}^n$.

A small detour: public-key encryption schemes. The foregoing description corresponds to the so-called model of a *private-key encryption scheme*, and requires the communicating parties to agree beforehand on a corresponding pair of encryption/decryption keys. This need is removed in *public-key encryption schemes*, envisioned by Diffie and Hellman [17] (and materialized by the RSA scheme of Rivest, Shamir, and Adleman [58]). In a public-key encryption scheme, the encryption-key can be publicized without harming the security of the plaintexts encrypted using it, allowing anybody to send encrypted messages to Party X by using the encryption-key publicized by Party X. But in such a case, as observed by Goldwasser and Micali [29], the need for randomized encryption is even more clear. Indeed, if a deterministic encryption algorithm is employed and the wire-tapper knows the encryption-key, then it can identify the plaintext in the case that the number of possibilities is small. In contrast, using a randomized encryption algorithm, the encryption of plaintext yes under a known encryption-key may be computationally indistinguishable from the encryption of the plaintext no under the same encryption-key. For further discussion of the security and construction of encryption schemes, the interested reader is referred to [23, Chap. 5].

E.2 Authenticated Communication

Message authentication is a task related to the setting discussed when motivating private-key encryption schemes. Again, there are two designated parties that wish to communicate over an insecure channel. This time, we consider an active adversary that is monitoring the channel and may alter the messages sent over it. The parties communicating through this insecure channel wish to authenticate the messages they send such that their counterpart can tell an original message (sent by the sender) from a modified one (i.e., modified by the adversary). Loosely speaking, a scheme for message authentication should satisfy the following:

- each of the communicating parties can *efficiently produce an authentication tag* to any message of its choice;
- each of the communicating parties can *efficiently verify* whether a given string is an authentication tag of a given message; but
- *it is infeasible for an external adversary* (i.e., a party other than the communicating parties) *to produce authentication tags* to messages not sent by the communicating parties.

Again, such a scheme consists of a randomized algorithm for selecting keys as well as algorithms for tagging messages and verifying the validity of tags.

A message authentication scheme based on pseudorandom functions. In the following message authentication scheme, a uniformly chosen n-bit key, s, is used for specifying a pseudorandom function (as in Definition 2.17). Using the key s, a plaintext $x \in \{0,1\}^n$ is authenticated by the tag $f_s(x)$, and verification of (x, y) with respect to the key s amounts to checking whether y equals $f_s(x)$. Again, the security of this scheme follows from the security of an imaginary (ideal) scheme in which f_s is replaced by a totally random function $F : \{0,1\}^n \to \{0,1\}^n$. For further discussion of message authentication schemes and the related notion of signature schemes, the interested reader is referred to [23, Chap. 6].

Appendix F

Some Basic Complexity Classes

This appendix presents definitions of most complexity classes mentioned in the primer (i.e., the time-complexity classes DTIME, BPTIME, $\mathcal{P}$, $\mathcal{BPP}$, $\mathcal{NP}$, $\mathcal{E}$, and $\mathcal{EXP}$ as well as the space-complexity classes DSPACE and $\mathcal{BPL}$). Needless to say, the appendix offers a very minimal discussion of these classes and the interested reader is referred to [24].

Complexity classes are sets of computational problems, where each class contains problems that can be solved with specific computational resources. To define a complexity class one specifies a model of computation, a complexity measure (like time or space), which is always measured as a function of the input length, and a bound on the complexity (of problems in the class).

The prevailing model of computation is that of Turing machines. This model captures the notion of algorithms. The two main complexity measures considered in the context of algorithms are the number of steps taken by the algorithm (i.e., its time complexity) and the amount of "memory" or "work-space" consumed by the computation (i.e., its space complexity).

P and NP. The class $\mathcal{P}$ consists of all decision problems that can be solved in (deterministic) polynomial-time. A decision problem S is in $\mathcal{NP}$ if there exists a polynomial p and a (deterministic) polynomial-time algorithm V such that the following two conditions hold:

1. For every $x \in S$ there exists $y \in \{0,1\}^{p(|x|)}$ such that $V(x,y) = 1$.
2. For every $x \notin S$ and every $y \in \{0,1\}^*$ it holds that $V(x,y) = 0$.

A string y satisfying Condition 1 is called an NP-witness (for x). Clearly, $\mathcal{P} \subseteq \mathcal{NP}$ and it is widely believed that the inclusion is strict; indeed, establishing this conjecture is the celebrated P-vs-NP Question.

Reductions and NP-completeness (NPC). A problem is $\mathcal{NP}$-complete if it is in $\mathcal{NP}$ and every problem in $\mathcal{NP}$ is polynomial-time reducible to it, where a polynomial-time reduction of problem Π to problem Π' is a polynomial-time algorithm that solves Π by making queries to a subroutine that solves problem Π' (such that the running-time of the subroutine is not counted in the algorithm's time complexity). Thus, any algorithm for an $\mathcal{NP}$-complete problem yields algorithms of similar time-complexity for *all* problems in $\mathcal{NP}$.

Typically, NP-completeness is defined while restricting the reduction to make a single query and output its answer. Such a reduction, called a Karp-reduction, is represented by a polynomial-time computable mapping that maps yes-instances of Π to yes-instances of Π' (and no-instances of Π to no-instances of Π'). Hundreds of NP-complete problems are listed in [19].

Probabilistic polynomial-time (BPP). A decision problem S is in $\mathcal{BPP}$ if there exists a probabilistic polynomial-time algorithm A such that the following two conditions hold:

1. For every $x \in S$ it holds that $\mathsf{Pr}[A(x)=1] \geq 2/3$.
2. For every $x \notin S$ it holds that $\mathsf{Pr}[A(x)=0] \geq 2/3$.

That is, the algorithm has two-sided error probability (of $1/3$), which can be further reduced by repetitions. We stress that due to the two-sided error probability of $\mathcal{BPP}$, it is not known whether or not $\mathcal{BPP}$ is contained in $\mathcal{NP}$. In contrast, for the corresponding one-sided error probability class, denoted $\mathcal{RP}$, it holds that $\mathcal{P} \subseteq \mathcal{RP} \subseteq \mathcal{BPP} \cap \mathcal{NP}$. Specifically, a decision problem S is in $\mathcal{RP}$ if there exists a probabilistic polynomial-time algorithm A such that (1) for every $x \in S$ it holds that $\mathsf{Pr}[A(x)=1] \geq 2/3$ whereas (2) for every $x \notin S$ it holds that $\mathsf{Pr}[A(x)=0] = 1$.

The exponential-time classes E and EXP. The classes $\mathcal{E}$ and $\mathcal{EXP}$ consist of all problems that can be solved (by a deterministic algorithm) in time $2^{O(n)}$ and $2^{\text{poly}(n)}$, respectively, for n-bit long inputs. Clearly, $\mathcal{NP} \subseteq \mathcal{EXP}$.

Generic time-complexity classes. In general, one may define a complexity class for every time bound and every type of machine (i.e., deterministic, and probabilistic), but polynomial and exponential bounds seem most natural and very robust. Indeed, for any time bound function $t : \mathbb{N} \to \mathbb{N}$, we may define the class $\text{DTIME}(t)$ (resp., $\text{BPTIME}(t)$) that consists of all problems that can be solved by a deterministic (resp., probabilistic) algorithm in time $t(n)$ for n-bit long inputs.

Space complexity classes. When defining space-complexity classes, one counts *only* the space consumed by the actual computation, and *not* the space occupied by the input and output. This is formalized by postulating that the input is read from a read-only device (resp., the output is written on a write-only device). Analogously to the generic time complexity classes, for any space bound function $s : \mathbb{N} \to \mathbb{N}$, we may define the class $\text{DSPACE}(s)$ that consists of all problems that can be solved by a deterministic algorithm in space $s(n)$ for n-bit long inputs.

We shall also consider the complexity class $\mathcal{BPL}$ that consists of all decision problems that are solvable by randomized algorithms of logarithmic space-complexity (and polynomial-time complexity). Thus, $\mathcal{BPL} \subseteq \mathcal{BPP}$.

We also mention the classes $\mathcal{L}$, $\mathcal{RL}$, and $\mathcal{NL}$, which are the logarithmic space-complexity analogues of $\mathcal{P}$, $\mathcal{RP}$, and $\mathcal{NP}$, respectively. Indeed, $\mathcal{L} \subseteq \mathcal{RL} \subseteq \mathcal{NL}$ holds (analogously to $\mathcal{P} \subseteq \mathcal{RP} \subseteq \mathcal{NP}$).

Bibliography

[1] M. Agrawal, N. Kayal, and N. Saxena. PRIMES is in P. *Annals of Mathematics*, Vol. 160 (2), pages 781–793, 2004.

[2] M. Ajtai, J. Komlos, E. Szemerédi. Deterministic Simulation in LogSpace. In *19th ACM Symposium on the Theory of Computing*, pages 132–140, 1987.

[3] R. Aleliunas, R.M. Karp, R.J. Lipton, L. Lovász and C. Rackoff. Random Walks, Universal Traversal Sequences, and the Complexity of Maze Problems. In *20th IEEE Symposium on Foundations of Computer Science*, pages 218–223, 1979.

[4] N. Alon, L. Babai and A. Itai. A Fast and Simple Randomized Algorithm for the Maximal Independent Set Problem. *J. of Algorithms*, Vol. 7, pages 567–583, 1986.

[5] N. Alon, O. Goldreich, J. Håstad, R. Peralta. Simple Constructions of Almost k-wise Independent Random Variables. *Journal of Random Structures and Algorithms*, Vol. 3, No. 3, pages 289–304, 1992. Preliminary version in *31st FOCS*, 1990.

[6] N. Alon and J.H. Spencer. *The Probabilistic Method.* John Wiley & Sons, Inc., 1992. Second edition, 2000.

[7] R. Armoni. On the Derandomization of Space-Bounded Computations. In the proceedings of *Random98*, Springer-Verlag, Lecture Notes in Computer Science (Vol. 1518), pages 49–57, 1998.

[8] H. Attiya and J. Welch. *Distributed Computing: Fundamentals, Simulations and Advanced Topics.* McGraw-Hill, 1998.

[9] L. Babai, L. Fortnow, N. Nisan and A. Wigderson. BPP has Subexponential Time Simulations Unless EXPTIME has Publishable Proofs. *Complexity Theory*, Vol. 3, pages 307–318, 1993.

[10] M. Bellare, O. Goldreich and M. Sudan. Free Bits, PCPs and Non-Approximability – Towards Tight Results. *SIAM Journal on Computing*, Vol. 27, No. 3, pages 804–915, 1998. Extended abstract in *36th FOCS*, 1995.

[11] M. Blum and S. Micali. How to Generate Cryptographically Strong Sequences of Pseudo-Random Bits. *SIAM Journal on Computing*, Vol. 13, pages 850–864, 1984. Preliminary version in *23rd FOCS*, 1982.

[12] M. Braverman. Poly-logarithmic Independence Fools AC0 Circuits. In *24th IEEE Conference on Computational Complexity*, pages 3–8, 2009.

[13] L. Carter and M. Wegman. Universal Hash Functions. *Journal of Computer and System Science*, Vol. 18, 1979, pages 143–154.

[14] G.J. Chaitin. On the Length of Programs for Computing Finite Binary Sequences. *Journal of the ACM*, Vol. 13, pages 547–570, 1966.

[15] B. Chor and O. Goldreich. On the Power of Two–Point Based Sampling. *Jour. of Complexity*, Vol. 5, pages 96–106, 1989. Preliminary version dates 1985.

[16] T.M. Cover and G.A. Thomas. *Elements of Information Theory*. John Wiley & Sons, Inc., New York, 1991.

[17] W. Diffie, and M.E. Hellman. New Directions in Cryptography. *IEEE Transactions on Information Theory*, IT-22 (Nov. 1976), pages 644–654.

[18] O. Gaber and Z. Galil. Explicit Constructions of Linear Size Superconcentrators. *Journal of Computer and System Science*, Vol. 22, pages 407–420, 1981.

[19] M.R. Garey and D.S. Johnson. *Computers and Intractability: A Guide to the Theory of NP-Completeness*. W.H. Freeman and Company, New York, 1979.

[20] O. Goldreich. A Note on Computational Indistinguishability. *Information Processing Letters*, Vol. 34, pages 277–281, May 1990.

[21] O. Goldreich. *Modern Cryptography, Probabilistic Proofs and Pseudorandomness*. Algorithms and Combinatorics series (Vol. 17), Springer, 1999.

[22] O. Goldreich. *Foundation of Cryptography: Basic Tools*. Cambridge University Press, 2001.

[23] O. Goldreich. *Foundation of Cryptography: Basic Applications*. Cambridge University Press, 2004.

[24] O. Goldreich. *Computational Complexity: A Conceptual Perspective*. Cambridge University Press, 2008.

[25] O. Goldreich, S. Goldwasser, and S. Micali. How to Construct Random Functions. *Journal of the ACM*, Vol. 33, No. 4, pages 792–807, 1986.

[26] O. Goldreich, S. Goldwasser, and A. Nussboim. On the Implementation of Huge Random Objects. In *44th IEEE Symposium on Foundations of Computer Science*, pages 68–79, 2003.

[27] O. Goldreich and L.A. Levin. Hard-core Predicates for any One-Way Function. In *21st ACM Symposium on the Theory of Computing*, pages 25–32, 1989.

[28] O. Goldreich and B. Meyer. Computational Indistinguishability – Algorithms vs. Circuits. *Theoretical Computer Science*, Vol. 191, pages 215–218, 1998. Preliminary version by Meyer in *Structure in Complexity Theory*, 1994.

[29] S. Goldwasser and S. Micali. Probabilistic Encryption. *Journal of Computer and System Science*, Vol. 28, No. 2, pages 270–299, 1984. Preliminary version in *14th STOC*, 1982.

[30] V. Guruswami, C. Umans, and S. Vadhan. Unbalanced Expanders and Randomness Extractors from Parvaresh-Vardy Codes. *Journal of the ACM*, Vol. 56 (4), Article No. 20, 2009. Preliminary version in *22nd CCC*, 2007.

[31] I. Haitner, O. Reingold, and S. Vadhan. Efficiency Improvements in Constructing Pseudorandom Generator from any One-way Function. In *42nd ACM Symposium on the Theory of Computing*, to appear.

[32] J. Håstad, R. Impagliazzo, L.A. Levin and M. Luby. A Pseudorandom Generator from any One-way Function. *SIAM Journal on Computing*, Volume 28, Number 4, pages 1364–1396, 1999. Preliminary versions by Impagliazzo *et al.* in *21st STOC* (1989) and Håstad in *22nd STOC* (1990).

[33] A. Healy. Randomness-Efficient Sampling within NC1. *Computational Complexity*, Vol. 17 (1), pages 3–37, 2008.

[34] R. Impagliazzo and A. Wigderson. P=BPP If E Requires Exponential Circuits: Derandomizing the XOR Lemma. In *29th ACM Symposium on the Theory of Computing*, pages 220–229, 1997.

[35] R. Impagliazzo and A. Wigderson. Randomness vs Time: Derandomization under a Uniform Assumption. *Journal of Computer and System Science*, Vol. 63 (4), pages 672-688, 2001.

[36] N. Kahale. Eigenvalues and Expansion of Regular Graphs. *Journal of the ACM*, Vol. 42 (5), pages 1091–1106, September 1995.

[37] D.E. Knuth. *The Art of Computer Programming*, Vol. 2 (*Seminumerical Algorithms*). Addison-Wesley Publishing Company, Inc., 1969 (first edition) and 1981 (second edition).

[38] A. Kolmogorov. Three Approaches to the Concept of "The Amount of Information". *Probl. of Inform. Transm.*, Vol. 1/1, 1965.

[39] E. Kushilevitz and N. Nisan. *Communication Complexity.* Cambridge University Press, 1996.

[40] F.T. Leighton. *Introduction to Parallel Algorithms and Architectures: Arrays, Trees, Hypercubes.* Morgan Kaufmann Publishers, San Mateo, CA, 1992.

[41] L.A. Levin. Randomness Conservation Inequalities: Information and Independence in Mathematical Theories. *Information and Control*, Vol. 61, pages 15–37, 1984.

[42] M. Li and P. Vitanyi. *An Introduction to Kolmogorov Complexity and its Applications.* Springer-Verlag, August 1993.

[43] A. Lubotzky, R. Phillips, and P. Sarnak. Ramanujan Graphs. *Combinatorica*, Vol. 8, pages 261–277, 1988.

[44] G.A. Margulis. Explicit Construction of Concentrators. *Prob. Per. Infor.*, Vol. 9 (4), pages 71–80, 1973 (in Russian). English translation in *Problems of Infor. Trans.*, pages 325–332, 1975.

[45] P.B. Miltersen and N.V. Vinodchandran. Derandomizing Arthur-Merlin Games using Hitting Sets. *Computational Complexity*, Vol. 14 (3), pages 256–279, 2005. Preliminary version in *40th FOCS*, 1999.

[46] M. Mitzenmacher and E. Upfal. *Probability and Computing: Randomized Algorithms and Probabilistic Analysis.* Cambridge University Press, 2005

[47] R. Motwani and P. Raghavan. *Randomized Algorithms.* Cambridge University Press, 1995.

[48] J. Naor and M. Naor. Small-bias Probability Spaces: Efficient Constructions and Applications. *SIAM Journal on Computing*, Vol. 22, 1993, pages 838–856. Preliminary version in *22nd STOC*, 1990.

[49] N. Nisan. Pseudorandom Bits for Constant Depth Circuits. *Combinatorica*, Vol. 11 (1), pages 63–70, 1991.

[50] N. Nisan. Pseudorandom Generators for Space Bounded Computation. *Combinatorica*, Vol. 12 (4), pages 449–461, 1992. Preliminary version in *22nd STOC*, 1990.

[51] N. Nisan. $\mathcal{RL} \subseteq \mathcal{SC}$. *Computational Complexity*, Vol. 4, pages 1-11, 1994. Preliminary version in *24th STOC*, 1992.

[52] N. Nisan and A. Wigderson. Hardness vs. Randomness. *Journal of Computer and System Science*, Vol. 49, No. 2, pages 149–167, 1994. Preliminary version in *29th FOCS*, 1988.

[53] N. Nisan and D. Zuckerman. Randomness is Linear in Space. *Journal of Computer and System Science*, Vol. 52 (1), pages 43–52, 1996. Preliminary version in *25th STOC*, 1993.

[54] N. Pippenger and M.J. Fischer. Relations Among Complexity Measures. *Journal of the ACM*, Vol. 26 (2), pages 361–381, 1979.

[55] A.R. Razborov and S. Rudich. Natural Proofs. *Journal of Computer and System Science*, Vol. 55 (1), pages 24–35, 1997. Preliminary version in *26th STOC*, 1994.

[56] O. Reingold. Undirected ST-Connectivity in Log-Space. In *37th ACM Symposium on the Theory of Computing*, pages 376–385, 2005.

[57] O. Reingold, S. Vadhan, and A. Wigderson. Entropy Waves, the Zig-Zag Graph Product, and New Constant-Degree Expanders and Extractors. *Annals of Mathematics*, Vol. 155 (1), pages 157–187, 2001. Preliminary version in *41st FOCS*, pages 3–13, 2000.

[58] R.L. Rivest, A. Shamir and L.M. Adleman. A Method for Obtaining Digital Signatures and Public Key Cryptosystems. *CACM*, Vol. 21, Feb. 1978, pages 120–126.

[59] M. Saks and S. Zhou. $\mathrm{BP_H SPACE}(S) \subseteq \mathrm{DSPACE}(S^{3/2})$. *Journal of Computer and System Science*, Vol. 58 (2), pages 376–403, 1999. Preliminary version in *36th FOCS*, 1995.

[60] J.T. Schwartz. Fast Probabilistic Algorithms for Verification of Polynomial Identities. *Journal of the ACM*, Vol. 27 (4), pages 701–717, October 1980.

[61] R. Shaltiel and C. Umans. Simple Extractors for All Min-Entropies and a New Pseudo-Random Generator. In *42nd IEEE Symposium on Foundations of Computer Science*, pages 648–657, 2001.

[62] R. Shaltiel. Recent Developments in Explicit Constructions of Extractors. In *Current Trends in Theoretical Computer Science: The Challenge of the New Century, Vol. 1: Algorithms and Complexity*, World Scientific, 2004. (Editors: G. Paun, G. Rozenberg and A. Salomaa.) Preliminary version in *Bulletin of the EATCS 77*, pages 67–95, 2002.

[63] C.E. Shannon. A Mathematical Theory of Communication. *Bell Sys. Tech. Jour.*, Vol. 27, pages 623–656, 1948.

[64] R.J. Solomonoff. A Formal Theory of Inductive Inference. *Information and Control*, Vol. 7/1, pages 1–22, 1964.

[65] L. Trevisan. Extractors and Pseudorandom Generators. *Journal of the ACM*, Vol. 48 (4), pages 860–879, 2001. Preliminary version in *31st STOC*, 1999.

[66] Y. Tzur. Notions of Weak Pseudorandomness and $\mathrm{GF}(2^n)$-Polynomials. Master Thesis, Weizmann Institute of Science, 2009. Available from the theses section of *ECCC*.

[67] C. Umans. Pseudo-random Generators for all Hardness. *Journal of Computer and System Science*, Vol. 67 (2), pages 419–440, 2003.

[68] S. Vadhan. *Lecture Notes for CS 225: Pseudorandomness*, Spring 2007. Available from `http://www.eecs.harvard.edu/~salil`.

[69] L.G. Valiant. A Theory of the Learnable. *CACM*, Vol. 27/11, pages 1134–1142, 1984.

[70] E. Viola. The Sum of d Small-Bias Generators Fools Polynomials of Degree d. *Computational Complexity*, Vol. 18 (2), pages 209–217, 2009. Preliminary version in *23rd CCC*, 2008.

[71] I. Wegener. *Branching Programs and Binary Decision Diagrams – Theory and Applications*. SIAM Monographs on Discrete Mathematics and Applications, 2000.

[72] A. Wigderson. The Amazing Power of Pairwise Independence. In *26th ACM Symposium on the Theory of Computing*, pages 645–647, 1994.

[73] A.C. Yao. Theory and Application of Trapdoor Functions. In *23rd IEEE Symposium on Foundations of Computer Science*, pages 80–91, 1982.

[74] R.E. Zippel. Probabilistic algorithms for sparse polynomials. In the *Proceedings of EUROSAM '79: International Symposium on Symbolic and Algebraic Manipulation*, E. Ng (Ed.), Lecture Notes in Computer Science (Vol. 72), pages 216–226, Springer, 1979.

Index

Titles in This Series